BRUNEL'S KINGDOM

BRUNEL'S KINGDOM

IN THE FOOTSTEPS OF BRITAIN'S GREATEST ENGINEER

JOHN CHRISTOPHER

The History Press

First published 2006
This edition published 2015
Reprinted 2017, 2019

The History Press
The Mill, Brimscombe Port
Stroud, Gloucestershire, GL5 2QG
www.thehistorypress.co.uk

British Library Cataloguing in Publication Data.
A catalogue record for this book is available from the British Library.

ISBN 978 0 7509 6306 0

Typesetting and origination by The History Press
Printed in Turkey by Imak

CONTENTS

INTRODUCTION

'Reader, if you seek a monument, look around you.'

When Isambard Kingdom Brunel (IKB) romped home to second place in the BBC's 2002 poll for the greatest Briton, it demonstrated a massive groundswell of popular enthusiasm for the man who shaped our modern world, a sentiment that was echoed by the 2006 bicentenary celebrations of his birth. During his lifetime, Brunel created some of the most sensational man-made structures in Britain: bridges that swept across rivers and gorges, railways that cut through the landscape and ships that shrank the world. These were the marvels of their age, marvels of any age for that matter, and each and every one has its own story to tell. And the most exciting thing is that so much of it is still out there and, in the main, accessible to the public.

The objective of *Brunel's Kingdom* is not only to celebrate Brunel's work but to encourage people to visit it. Not in a museumy sort of way and not just the big signature stuff, the showpiece landmarks, but also the many lesser-known examples. Using this book as an inspiration and guide, I would encourage readers to get out there and touch it, walk or ride on it, photograph it and most importantly go and experience it for themselves! Coming face to face with these spectacular monuments can be as exciting as meeting a famous film star, and sometimes there is the shock of discovering that reality actually matches the history books.

The main text of this book provides the historical context and background to Brunel's work in the form of a journey arranged by geographical order, although incidentally, mostly chronological too. It starts in London with the Thames Tunnel and moves westwards along the Great Western Railway (GWR) stopping at Swindon, Bath, Bristol and other towns before heading on to Devon for the Atmospheric Railway, the spectacular Royal Albert Bridge over the Tamar at Saltash and the sites of the wooden viaducts in Cornwall. You could say that the journey then continues across the Atlantic to New York before coming full circle back to London. Each section is accompanied by a guide telling you how to find the sites with, where appropriate, opening times and so on, as well as sources of further information. On pages 143–4, suggestions are made for grouping locations into longer round trips.

The quotation at the start of this introduction is translated from the Latin inscription to Sir Christopher Wren at St Paul's Cathedral. What better advice could you ask for? Brunel's work is his greatest monument, so go explore!

The figure of Brunel
at the Swindon
Steam museum.

USING THE GUIDE

For those equipped with street maps for major cities and some towns, or Ordnance Survey (OS) Landranger 1:50,000 scale maps for smaller towns and rural areas, locating sites mentioned in this book should be straightforward. In some instances the locations themselves – or the land providing access to them – are privately owned and not necessarily open to the public. Access to railway sites is often restricted, either by the nature of their location or on the grounds of safety, and the rail companies threaten to prosecute trespassers. Please note that while every effort has been made to provide the most up-to-date information for this guide, details can change. This may apply to museums and public attractions, so check on opening times and so on before setting off.

Brunel's Kingdom was originally published in 2006, and this new 2015 edition has been extensively updated.

Brunel
photographed
against the
chains at
Millwall, 1859.

ACKNOWLEDGEMENTS

Producing *Brunel's Kingdom* has only been possible with the kind help and assistance of a great many people and my thanks go to all of them, including:

Robert Hulse of the Brunel Museum at the Brunel Engine House; Mark Middleton and Catherine Langford of Nicholas Grimshaw Partners; Dr Steven Brindle, English Heritage; Felicity Ball, Curator of Steam, the Museum of the GWR; John Mitchell and Mike Rowland, Clifton Suspension Bridge Trust; Andy King, Curator of Bristol Industrial Museum; the SS *Great Britain* Trust; Felicity Cole from the Newton Abbott Museum; and Amy Rigg of The History Press for commissioning this revised edition. Special thanks to Ute, who kept me and this book on track, and to Anna and Jay, who have learnt so much about 'Mr Brunel', as they call him, and have joined me on many of my expeditions.

All new photography is by John Christopher unless otherwise credited.

A plaque indicating the position of No.1 Britain Street in Portsmouth, the house in which Brunel was born on 9 April 1806.

1

THE THAMES TUNNEL

Rotherhithe, London

'Ladies and gentlemen, welcome to the eighth wonder of the world!'

The bottom of an escalator in Rotherhithe tube station doesn't exactly tally with a claim that this place ranks alongside the pyramids of Egypt or the Hanging Gardens of Babylon. So who could blame the other Saturday afternoon travellers for scurrying past with worried frowns? Had they stumbled upon a gathering of some obscure sect that worships at the secret subterranean shrines of London transport? Not that there was anything remotely secretive about our guide's enthusiastic delivery over the background clatter of people and trains. I had braved London on the last weekend before Christmas, not to indulge in a little last-minute shopping, but to see a tunnel. A strange thing to visit, a hole in the ground normally glimpsed from a moving train as it flashes through with the faint rhythmic 'whumpf, whumpf' sound of air beating against the arches. Then it's gone … unless, that is, you get yourself on to one of the tours organised by the Brunel Museum. Situated on the south side of the Thames, just an energetic stone's throw from Rotherhithe station, this museum has been established by a band of enthusiasts to celebrate the achievements of two outstanding engineers.

Of all the tunnels in the world, what makes this one so special? Fortunately Robert Hulse, curator at the Brunel Museum, was on hand

to explain as we gathered beneath an impressive plaque. Now, I have to confess that I have mixed feelings regarding these designer labels for buildings, but this one is a beauty. 'This is an International Landmark Site plaque,' Robert announced. 'There are only six of these in the country, and this is the most important … although I would say that. There are 200 of these plaques around the world, and I am told by civil engineers that this site is in the top ten. It is not just the oldest section of tunnel in the London Underground, not just the first project that Isambard Kingdom Brunel worked on and the only one that father and son worked on together. This is where it *all* began.'

And by *all* he doesn't only mean *all* underground railways. He's talking about the extraordinary career of I.K. Brunel in particular. This, then, is where our journey in his footsteps must begin.

Appropriately, the plaque was erected jointly by the American Society of Civil Engineers and the British Institution of Civil Engineers. Although Brunel senior was French by birth, he became an American citizen and died British. Born in 1769, Marc Isambard Brunel had evaded a life in the church to join the navy. As a staunch royalist, he had been forced to flee the 'Terror' of the French Revolution at the age of twenty-four by setting sail for the New World, the land of opportunity. Once there, Marc quickly demonstrated a flair for applying his inventive mind to a wide range of disciplines including surveying, architecture and engineering. As Robert explained:

> He was appointed as Chief Engineer of the Port of New York and went on to advise on the defences on Long Island and Staten Island. He made plans for the first canal in North America, from Lake Champlain up to the Hudson, and he won

a competition for the design of the new Congress buildings in Washington – although the buildings were never put up as they were too expensive. I think this was a family trait. The Brunels designed wonderful stuff, but it didn't half cost!

Despite doing well, Marc abandoned his new life and sailed to England because he was deeply in love. While still in France, he had met a young English governess named Sophia Kingdom, and after six long years apart they were reunited in March 1799 and married before the year was out. They later settled in Portsmouth and had three children: Sophia and Emma, and then Isambard, who was born on 9 April 1806. His first name comes from his father (also known as Isambard, although widely referred to as Marc to avoid confusion) and Kingdom is from his mother's side. For Marc Brunel there was absolutely no doubt, no question whatsoever, that his one and only son would become a great engineer.

Love aside, there was an altogether more prosaic reason for Marc's arrival on British shores. He had learnt of the Admiralty's urgent need to improve the production of ship's pulley blocks – an item of equipment which might not appear significant until you consider that one large ship of the line needed approximately 1,400 pulleys. An expansion of the fleet required huge quantities, but the blocks were being laboriously produced by hand by skilled craftsmen. Marc was about to change all that with his revolutionary methods for mass production. Armed with sheaves of 'mechanical drawings' – a technique for accurately portraying three-dimensional objects still largely unknown outside of French military circles – he teamed up with Henry Maudsley in London, who was probably the greatest mechanic of his time. Although their proposals were met with resistance from the established block-makers – who were doing very nicely, thank you – they came to the attention of the Inspector-General of Naval Works and, as a result, a new block-making plant was commissioned for Portsmouth. With Maudsley constructing the machines under Marc's supervision, their integrated system of forty-three machines slashed the workforce requirement by 90 per cent and, in the first year of operation, saved the Admiralty a whopping £24,000. Marc received £17,000 for his work, a considerable sum (although less than he anticipated), and the family moved to a bigger house in Chelsea.

Then began a period of mixed fortunes for Marc Brunel. In addition to investing capital in a sawmill at Battersea, which appears to have been

At Rotherhithe, the capped top of the circular shaft is clearly visible to the left of the Brunel Engine House. The roof of the Underground station is at the top of the picture. (The Brunel Museum)

profitable, he applied himself to a number of projects ranging from dock improvement schemes, bridge designs and even a plan for a Panama canal, to devices for winding cotton and making ornamental tinfoil. He also won an Admiralty contract to improve timber handling and sawing at the Chatham Dockyard that, incidentally, included a tunnel and a 'wide-gauge' railway for moving the timber. And it was at Chatham that Marc stumbled upon a new idea while observing a lowly woodworm! He noted that the *teredo navalis* was able to bore its way through ships' timbers because it was protected by hard shells on either side of its head and excreted the waste to line its own tunnel. From this he formulated his 1818 patent for a device for 'Forming Drifts and Tunnels Underground' – in other words, a tunnelling shield. In its original form it consisted of a cylinder divided into six compartments in which the miners worked, capable of edging forward independently of each other. Wooden boards or 'pollings' secured the exposed workface, and when each compartment had been excavated to its full extent, hydraulic jacks thrust the device forward, exposing a series of cast-iron rings at the back to line the tunnel. In essence this was a blueprint for the cylindrical tunnelling shields still in use today.

Meanwhile, Marc's other enterprises were not going well. Extensive experimentation with a 'Gaz Engine' fuelled by liquefied gas came to nought, and other schemes were plagued by bad luck or poor timing. For example, take his mechanical process for the mass production of footwear. Boots were desperately needed for the British troops fighting in France, and encouraged by the government, Marc established a factory manned by war veterans to churn out a staggering 400 pairs per day. Unfortunately, in the aftermath of Napoleon's defeat, this prodigious output quickly piled up into an enormous unwanted boot mountain. In fact, Marc's financial affairs were in such poor order by 1821 that he and Sophia endured a spell in the debtors' prison. It was only thanks to their influential friends and the direct intervention of the Duke of Wellington that they were released. This enforced period of introspection must have given Marc time to evaluate his many unrealised ideas, including the tunnelling shield. The concept was sound; all he needed was a tunnel to build.

Tunnel Vision

At the dawn of the nineteenth century, Britain was painting the world map red with empire, resulting in a massive increase in the merchant shipping pouring into the Port of London. With each ship required to berth at 'legal quays', the throng of vessels vying for space soon became unmanageable and new docks were planned for the Isle of Dogs, at Wapping and Blackwall as well as on the south side of the Thames. As a direct result, this added to the daily congestion on the roads, with thousands of wagons heading for the nearest river crossing, the old London Bridge. Londoners may not have heard of 'gridlock' back then, but they certainly knew what it meant. One solution might have been to build another bridge nearer the docks, but it would have to be

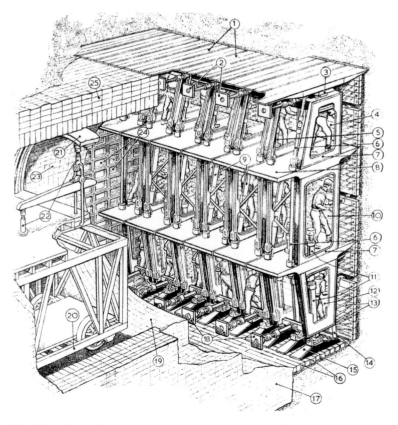

Marc Brunel's tunnelling shield. Each vertical frame accommodated three compartments for the miners, one above the other. The workface, shown on the right, is held in place by polling boards, each one positioned by a pair of screw jacks. Once the excavation for a frame was completed with all the boards extended forward, the frame itself was repositioned by hydraulic jacks pushing against the completed brickwork at the back. (The Brunel Museum)

high enough for tall-masted ships to pass underneath. The alternative, a 'bascule' or lifting bridge, was impossible as the current steam engines were not up to the job, and it is quite likely that the frequency of the shipping meant such a bridge would remain almost constantly raised. Perhaps a tunnel under the river was the answer? The idea had been mooted as early as 1798, but the ground beneath the Thames consists mostly of clay, which is fine for tunnelling, but also layers of unstable gravels and even quicksand. Consequently the construction of a sub-aqueous tunnel was widely deemed to be impossible. To an engineer, however, 'impossible' is merely another word for 'challenging'!

One such individual was Robert Vazie, a Cornish tin mining engineer with some experience of driving mine shafts under the coast. Vazie's plan was to dig a narrow timber-framed pilot tunnel, known as a 'driftway', which would later act as a drain for a full-sized tunnel. In July 1805 he started excavating a vertical shaft at Rotherhithe on behalf of the newly formed Thames Archway Co., but the works were soon overwhelmed by water and the money ran out. More capital was raised and the shaft continued downwards for 76 feet, where quicksand was encountered. At this point the company brought in another Cornishman to take charge, the renowned engineer Richard Trevithick. Under his direction the driftway had actually been driven 1,000 feet under the river, with only 200 feet to go, when the water broke in. Trevithick and his men escaped with their lives, but he was unable to raise enough backing to complete the tunnel workings and they were abandoned. Marc Brunel must have followed these developments closely as he had already considered a tunnel under the River Neva, at St Petersburg, on behalf of the Russian Tsar – plans which were never implemented. Now he had an opportunity closer to home.

Marc set about modifying the design of the shield from the original cylinder to a rectangle because he wanted to raise the stakes from Vazie's modest driftway to a twin tunnel, 38 feet wide and 22 feet 6 inches high. Twelve vertical cast-iron frames, each 3 feet wide, formed the new shield. The frames were divided vertically into three cells – a space roughly equivalent to the inside of a telephone box – with a miner working in each. Facing a series of horizontal polling boards held in place by screw jacks, the miner would remove one board at a time and excavate a few inches before placing that board against the new face. Each vertical frame sat on a movable iron foot connected via a ball joint and, once all boards had been repositioned, it was shunted forward by hydraulic jacks pushing against the brickwork behind. Pivot plates

known as 'staves' protected the top of the shield, and at either side iron plates were repositioned as it edged forward.

Having gathered together his backers, Marc Brunel began work on the Thames Tunnel in 1825 – the same year that the Stockton and Darlington Railway began putting steam railways on the map. George IV was King of England, to be followed by William IV, and it would be another twelve years until Queen Victoria ascended to the throne. All the wonders of the modern age that we take so much for granted – photography, telephones, wireless, electric lighting and the internal combustion engine, for example – had yet to be invented. Charles Dickens was only thirteen years old, the Duke of Wellington and Marc Brunel were both fifty-six, and working at Marc's side was his nineteen-year-old son, Isambard.

Marc decided to drive the tunnel between Rotherhithe and Wapping – a distance of 1,200 feet and only about 14 feet or so below the riverbed's lowest point. But before he could excavate the tunnel, he needed to dig an access shaft. A huge 50-foot diameter cylinder, 42 feet high, was constructed above ground at Rotherhithe, just three quarters of a mile from Trevithick's abandoned driftway. Built of brickwork, it was 3 feet thick and braced with vertical metal rods, which were secured to iron rims top and bottom. This brick tube worked much like a massive pastry cutter, as it sank under its own weight at a rate of about 6 inches per day, while the ground was excavated beneath it. This was such a marvel in its own right that visitors flocked from all over London to observe its progress. Once the full extent of the shaft was below ground, further excavations were made to underpin it and to accommodate a drainage reservoir. Then, in November 1825, the tunnelling shield, built by Maudsley, was positioned at the bottom of the shaft to start boring the tunnel.

From the platform of Rotherhithe Underground station nowadays, you can't make out Brunel's tunnel or the vertical shaft, as you can't see beyond the blackened tube formed by the railway's later extensions. This is why I had joined Robert Hulse's tour group as it boarded a train. Special arrangements had been made with London Underground to keep the maintenance lights switched on and our driver took us through at a snail's pace. As we pulled away we passed directly beneath the shaft, and then the newer brickwork finished and the long string of interconnecting arches between the original twin tunnels appeared. The first four arches have been left unrendered to reveal the original bricks, now dark with age. A young woman in our group turned excitedly to her companion, 'To think this is Brunel's brickwork!' I smiled to myself, but you can't help sharing her awe. Here's history at your fingertips. The remaining sixty arches, with their elegant Doric columns, have been rendered in concrete to protect them.

Cocooned by the relative comfort of a tube train, it is impossible to imagine the conditions in which the tunnellers must have laboured. For the main part, the workmen had been recruited from Cornish tin mines, or from the Durham coal mines in the North East. This workforce had its own hierarchy, with the chief engineer and assistant engineers ('gentlemen') at the top, the foremen – not quite 'gentlemen' but 'commanders of men' – and the 'working people', the miners, bricklayers and labourers, at the bottom. The wages for both skilled and unskilled craftsmen were perhaps a little better than for most workers in London at that time. In addition they were paid in kind with food, sweet porter ale or spirits, and a supply of dry clothes. Even so, for gentleman and labourer alike, the conditions were truly appalling. The Thames was nothing more than an open sewer serving a city of over two million people. Filthy water seeped in through the shield and often the miners in the lower frames were working in it up to their knees. The men retched in the foul air, their fingernails rotted and their skin was covered with sores. Many were struck down by 'tunnel sickness', a condition which sometimes caused blindness, and one man ended up in an asylum. Another danger was constantly present in the form of pockets of inflammable methane gas, which would suddenly ignite into balls of flame.

Operating in eight-hour shifts – sometimes less if the fumes were too noxious – thirty-six miners worked within the shield. Behind them were the bricklayers, each laying up to 1,000 bricks a day, and then the labourers scurrying up and down the tunnel removing the spoil, mixing cement and carrying materials. Inevitably the terrible conditions took their toll and by April 1826 the original resident engineer, William Armstrong, had resigned. For Marc Brunel, also ill by this time, there was only one candidate to take over the task and so Isambard became resident engineer, albeit on an acting basis initially. Undaunted by the enormous responsibility, he threw his heart and soul into the project and for the next two years he ate, breathed and slept Thames Tunnel.

Portraits of the young engineer show an unremarkable face with a long nose, full lips and a hairline already receding a little. But if you could have looked into those alert eyes, you would have seen a spark of fire, a man driven by the endless opportunities open to him at the dawn of an industrial and transportation revolution that would reshape the world. It was his destiny. It was what he had been born for. In our own more cynical age it is hard to understand that someone could devote their life to their job, even give their life to it, but for Isambard Kingdom Brunel it was the natural thing to do. Understandably Marc was proud of his son, and he wrote in his journal:

Isambard's vigilance is of great benefit. He is in every respect a most useful coadjutor in the undertaking … Isambard was the greater part of the night in the works and the benefit of his exertions is indeed most highly felt: no one stood out like him.

Isambard spent more than twenty hours a day in the tunnel with only catnaps to keep him going, often for days on end. With Marc in overall control and Isambard driving the work on site, progress was good and by the beginning of 1827 they had driven the tunnel 540 feet – almost halfway to Wapping. However, both men were increasingly anxious about the danger of water breaking through. The riverbed was far too close for comfort and Marc recorded his fears:

A work that requires such close attention, so much ingenuity and carried on day and night by the rudest hands possible – what anxiety, what fatigue, both of mind and body … Every morning I say, Another day of danger over.

His concerns weren't only for the workforce. The company directors were admitting a stream of shilling-sightseers to witness the great work close at hand, with up to 700 people entering the works daily. It is hard to conceive any circumstance nowadays where such a high-risk operation would open its doors to the paying public. Besides, the Brunels had good cause to be concerned. Instead of a seam of clay, the tunnellers encountered increasing amounts of gravel, often with debris such as bits of china, wood or a shoe buckle and other detritus falling through. Marc estimated that at high water the shield was supporting maybe 600 tons of water, and it was only by sheer good fortune that the first major deluge occurred at a time when no visitors were present.

Flooded!

On the evening of 18 May 1827, Isambard was further back in the tunnel while his assistant, Richard Beamish, worked at the shield where frames No.9 and No.11 were about to be moved forward. All day there had been more than the usual amount of seepage. Then, suddenly, No.11 erupted with jets of water! Within moments a thunderous torrent burst through and Isambard looked up to see Beamish and the panic-stricken workmen running for their lives. A small office building erected within the tunnel was flattened by the surge and the lights flickered out, leaving nothing but blackness and the deafening noise. Isambard shouted over the mayhem, ordering the men to get to the shaft as quickly as possible, but as they neared its top some of the lower steps were swept away. Suddenly a cry of help was heard from below, and tying a rope around his waist, Isambard slid down one of the iron ties of the shaft to rescue an old engineman named Tillett. Unbelievably there had been no casualties that night and Marc Brunel felt an enormous sense of relief; the inevitable had finally happened and they had got away with it.

For Isambard, his adrenalin-fuelled escape had been a heady and enticing cocktail. Now he borrowed a diving bell from the West India Dock Co. and descended beneath the cold waters of the Thames to investigate the damage. On the riverbed he found a depression, probably caused by gravel dredgers, and he was able to place one foot on the top of the No.12 frame of the shield. From his journals we glimpse a figure more akin to Indiana Jones than the familiar stern Victorian in a stovepipe hat:

What a dream it appears to me! Going down in the diving bell, finding No.12! The novelty of the thing, the excitement of the occasional risk attending our submarine excursions, the crowds of boats to witness our works all amused …

In order to plug the hole, 150 tons of clay, packed in bags and reinforced with hazel twigs, was dumped in the river and the pumps began to clear the tunnel. By 13 June, Isambard was able to inspect it by boat. What is not shown in many contemporary engravings is that the tunnel actually dips in the middle by 6–7 feet – although you can see this from Wapping Underground station – and so as the boat party went further in, they were able to propel themselves by pushing against the roof until grounded by the mound of silt burying the shield. Thankfully it was still intact and then began the long, laborious job of clearing the mess.

This story has a strange sequel, once the workings had been cleared. As Robert Hulse explained:

Just before Christmas 1827 the Brunels held the most extraordinary dinner party of all time. The tunnel was draped with red damask hangings, they had tables with white silk cloths, solid silver place settings and crystal glass. The Duke of Wellington was there and many notables attended the world's first underwater banquet!

To entertain the dinner guests, the Band of the Coldstream Guards played 'Rule Britannia!', the National Anthem, 'See the Conquering Hero Comes' – which was Wellington's theme tune whenever he turned out – and for a little light relief, an aria by Weber. Wonderful stuff! But it must have been deafening. Accordingly, at this point one of the guests went up to the band saying, 'Well that was awfully nice, but don't you chaps think you would like a little rest now?' Within the adjacent archway, 150 of the workmen were also entertained and as a mark of their great respect for Isambard, they ceremoniously presented him with a pick and shovel.

There was to be no rest for the workmen, however, and once tunnelling resumed it was clear that the conditions were as treacherous

A banquet beneath the Thames! To celebrate successfully clearing out the flooded tunnel in 1827, dinner guests were entertained by the Band of the Coldstream Guards. The figure of Marc Brunel can be seen standing front left. (Ironbridge Gorge Museum Trust, Elton Collection)

as ever. Determined to make progress, Isambard believed that if they pushed on quickly enough they would find more stable ground, but this strategy had tragic consequences. Early on the morning of 12 January 1828, disaster struck again. 'The roar of the water in a confined space was very grand, cannon can be nothing to it,' Isambard later recalled. Typically, he had been working alongside the miners for most of the night:

At six o'clock in the morning a fresh set or shift of the men came on to work. We began to work the ground at the west top corner of the frame; the tide had just began to flow, and finding the ground tolerably quiet, we proceeded by beginning at the top, and had worked about a foot downwards, when on exposing the next six inches the ground swelled suddenly, and a large quantity burst through the opening thus made. This was followed instantly by a large body of water.

The torrent was so violent that the man working that cell was thrown backwards on to the timber staging. Isambard, in the same frame, managed to clamber into the neighbouring compartment and tried to stem the flow: 'Seeing that there was no possibility of then opposing the water, I ordered all the men in the frames to retire.' He set off along the west arch with the last of them when they were plunged into darkness. The freezing water was already up to their waists and impeding their progress:

I was at that moment giving directions to the three men in what manner they ought to proceed in the dark to effect their escape, when they and I were knocked down, and covered with part of the timber stage. I struggled under water for some time, and at length extricated myself from the stage, and by swimming and being forced by the water, I gained the eastern arch, where I got a better footing, and was enabled, by laying hold of the railway rope, to pause a little, in the hope of encouraging the men who had been knocked down at the same time with myself.

The deluge continued and by the time he reached the shaft, Isambard was swimming for his life, struggling to reach the visitors' stairs: 'My knee was so injured by the timber stage that I could scarcely swim or get up the stairs, but the rush of water carried me up the shaft.' In fact, the

force of water was so strong that it rose right up to the lip of the shaft and the semi-conscious Isambard was plucked to safety by the others. He survived, but six others did not, including four who had reached the ladders only to be sucked back by the column of water as it receded.

Isambard had sustained serious injuries, both to his leg and internally, but even so he called for the diving bell and directed operations from a mattress on the deck of the barge. Subsequent inspections revealed that the damage to the tunnel was far greater than with the first breach, requiring thirty times more clay to fill the hole. By May, the remaining funding was dwindling away and the directors ordered all work to be suspended. They sent in bricklayers to seal the workface. 'Saw the last of the frames!!!' wrote an exasperated Marc Brunel in August 1828. As a final ignominy, a large mirror was placed against the brick wall and visitors continued to pay to see what *The Times* disparagingly referred to as 'the Great Bore'.

Isambard was sent to Brighton and afterwards on to Bristol to recuperate, and he was destined to never work on the Thames Tunnel again. Some efforts were made to reawaken interest but by the end of 1831 it seemed hopeless and, despondently, he wrote:

Tunnel is now, I think, dead … This is the first time I have felt able to cry at last for these ten years. Some further attempts may be made – but – it will never be finished now in my father's lifetime I fear. However, *nil desperandum* has always been my motto – we may succeed yet …

Yet despite these bleak predictions, there was some light at the end of this tunnel. Thanks to a Treasury loan of £270,000, work resumed in 1835 with an improved shield to replace the bricked-up and rusty original. In spite of further floods the tunnel was finally completed, albeit sixteen long years after its commencement. It officially opened to the public on 25 March 1843, and within fifteen weeks had received one million visitors. Queen Victoria rewarded Marc with a knighthood and she even turned up at Wapping in the royal barge for a personal tour. For a while, the Thames Tunnel truly was the eighth wonder of the world!

Unfortunately, admiration and high visitor numbers were not enough to bring financial success; during Sir Marc's lifetime the tunnel was never used for its intended purpose – to move cargo – and for a very good reason. They could not persuade horses to go down the stairs! As

Robert explains: 'They were going to build two further shafts four times the size of the existing ones, with a ramp for the horses and wagons to take cargo down to the tunnel, and then they would have made money.' Without these ramps the tunnel was only fit for pedestrians paying a penny a head, and this didn't even cover interest on the bank loan. 'So they had to think of other ways of getting money off people before they let them out, and today we know this as a *retail opportunity*. It became the world's first underwater shopping precinct, and in each of the sixty-four arches a shop-keeper would sell their wares to the people coming and going.'

There are many examples of this branded merchandise in the museum – Thames Tunnel gin flasks, Thames Tunnel pin cushions, spectacle cases, cigar-cutters, even nurseryware: 'If they had had baseball caps they would have put the Thames Tunnel on them and they'd have sold them!' Unfortunately, it was also discovered that there were other ways to get money off people and the tunnel became the haunt of thieves and prostitutes. The desperate directors launched the first underwater

Then: This contemporary view of the Thames Tunnel reveals an elegant structure, although typically the scale has been somewhat exaggerated. But where is the horse-drawn traffic for which the tunnel was built?

Now: The entrance of the tunnel as it is today, seen from the platform at Wapping. Despite the railway line passing through, the twin horseshoes still look magnificent.

The interior of the Wapping shaft. A modern lift runs up its centre, linking the Underground platforms with street level, but take the stairs to get a real impression of its scale and to see the original brickwork.

fairground in 1852 – The Thames Tunnel Annual Fancy Fair – with acts including serenaders, dancers, Electricity the Great American Wizard, and Mr E. Green, the celebrated bottle pantomimic equilibrist!

In 1865, the Thames Tunnel was adapted for rail travel when it was sold to the East London Railway for £200,000, opening to the public four years later, and it has remained an important transport route ever

since. When our tour group's train reached Wapping, I discovered that it has the narrowest platforms I have ever seen. The station walls are decorated with engravings of the tunnel during its construction and working life. But, best of all, from the end of the platforms you can see the original inverted twin horseshoe entrances. It's like looking down the barrels of a gigantic shotgun, although the dip means you can't see all the way through. Turning up a short flight of stairs at the end of the platform, you enter the bottom of the Wapping shaft, which has a lift running up its centre to street level. Take the spiral stairs hugging the original brickwork to get a real sense of this 'underground tower'. They lead up to the white-tiled booking hall, where there is another plaque honouring the Brunels. The top of this shaft has been incorporated within the structure of the modern station, which huddles amid tall dockside buildings with evocative names such as Gun Wharves, King Henry's Wharves and New Crane Wharf.

Following completion of the first phase of the extension of the East London Line, in 2010, the tunnel has officially become part of the new London Overground, albeit passing under the Thames.

The Brunel Engine House

From Wapping I made a quick hop back through the 'time tunnel' to Rotherhithe to investigate the Brunel Museum at the Brunel Engine House. Out of Rotherhithe station, the street names are proof enough that you are in the right place – Brunel Road, Railway Road, Tunnel Road. The restored chimney of the Brunel Engine House is easily spotted among the historic dock buildings, cheek-by-jowl with new riverside apartments and old London housing stock. St Mary's church is just around the corner, and from the riverside you can spot Gun Wharves over on the Wapping side of the river.

The Engine House is a red-brick building about 40 feet by 20 feet, built by Marc to house a steam engine to pump water out of the tunnel. Electric pumps took over in 1913, and after that the building was let to various tenants as a shed and lastly as a stonemasons' premises until it became a roofless ruin. A charitable trust was formed to save it in 1973, and it was first opened as a museum in 1980. Beside it is the capped top of the Rotherhithe shaft, which is surmounted by a high fence and painted in a very non-industrial light blue, making it look more like a water tank. Both shaft and Engine House are peppered with plaques, but I was more interested in seeing inside and once again Robert was on hand.

On the upper balcony there is a display of nineteenth-century merchandise, plus excellent models of the tunnel and shield. Dominating the central display space is an imposing horizontal compound steam engine painted deep green and highlighted with red detailing and dark wooden cladding. Built by J. & G. Rennie of Southwark, it was originally used in Chatham Dockyard in the 1880s. Rennie had won the contract to build the second shield after works reopened on the tunnel in 1835 and also built a pumping engine for the Wapping shaft. If the engine ran on steam, then this place runs on pure enthusiasm. When Robert moved to the area, he came upon it quite by chance and became involved as a volunteer:

> With the support of the Board of Trustees I have tried to open up the site to a wider audience, especially children. You come

at a very exciting time because we have just been awarded a Heritage Lottery Grant to fund education and outreach work, and another grant from Renaissance London to further improve the exhibition.

In addition the Brunel Museum organises tours by train and on foot through the tunnel every year, as well as a programme of school visits, festivals and community arts events. The next ambitious step is to convert the empty shaft into an exciting exhibition area right above the operating railway!

'Bristol has some wonderful examples of IKB's work, but we like to remind people that his career began and ended in London,' says Robert. However, before I headed westwards, there was another Brunel connection to follow up – this time over the Thames.

The great brick tube at Rotherhithe, which acted like a giant pastry cutter to create an access shaft as the soil was excavated.

THE THAMES TUNNEL

LOCATION: Between Rotherhithe and Wapping.

GETTING THERE: East London Line.

The tunnel is part of a busy railway line, but special tours are arranged by the Brunel Museum.

Visit Wapping for the twin horseshoe entrances. Access between platforms and street level is via the original shaft – so take the stairs! Check out the plaque erected by London Transport in 1959 in the booking hall.

You can't see much of the tunnel from Rotherhithe station as the brickwork nearest to the platforms is a later extension. Above the stairs on the way out is the International Landmark Site plaque.

THE BRUNEL MUSEUM

LOCATION: Brunel Engine House, Railway Avenue, Rotherhithe, SE16 4LF.

GETTING THERE: Underground, East London Line to Rotherhithe.

OPENING TIMES: Daily 10.00–17.00.

On-street parking in the area is reasonably good – it is outside the Central London Congestion Zone. From Rotherhithe station turn left, and left again towards the river and the tall dark chimney of the Engine House along Railway Avenue. Next to it is the capped top of the Rotherhithe shaft. Plaques galore.

INFORMATION: 020 7231 3840 / www.brunelenginehouse.org.uk

PORTSMOUTH BIRTHPLACE

The house where IKB was born in Britain Street, Portsmouth, is no longer there – a plaque marks the location a short walk from HMS *Victory*. (www.hms-victory.com)

INFORMATION: www.memorials.inportsmouth.co.uk/dockyard/Brunel.htm

CHATHAM DOCKYARD

Marc Brunel's sawmill building survives at the former dockyard at Chatham, Kent. This closed as a naval base in 1984 and the 80-acre site is now a huge museum complex, including the Museum of the Royal Dockyard. (www.thedockyard.co.uk)

BRUNEL'S HOMES IN LONDON

IKB's childhood home from 1811–26 is part of the seventeenth-century Lindsey House in Chelsea. Known then as 4 Lindsey Row, now 98 Cheyne Row. Owned by the National Trust, this is a private house not open to the public. A fifteen-minute walk from South Kensington station.

The site of his later home and offices at 18 Duke Street, Whitehall, is now occupied by government buildings.

Opposite: A plaque at the Engine House in Rotherhithe.

BRUNEL'S ENGINE HOUSE

★ ★ ★

The tunnel shaft and pumping house for Marc Brunel's tunnel was constructed between 1825 and 1843. This was the first thoroughfare under a navigable river in the world.

2

HUNGERFORD BRIDGE

London: The Missing Bridge

It is a conundrum worthy of Lewis Carroll. Where can you find Brunel's suspension bridge in full sight and yet hidden? It was one and is now three. And it could be said to span two rivers, a hundred miles apart.

Coming from the Thames Tunnel, I had travelled a couple of miles west to emerge at Temple station on the Victoria Embankment. I was on the trail of the missing bridge – or the missing link if you prefer – and I found the first clue at the end of Temple Place, where stands one of the finest statues of Isambard Kingdom Brunel in the country. It is an informal and bare-headed figure (minus his trademark stovepipe hat), ebony black and raised high to overlook the river – although originally it had been intended for Parliament Square. It was sculpted by Baron Marochetti and erected in 1877 by the Institution of Civil Engineers. The concise inscription reads, 'Isambard Kingdom Brunel. Civil Engineer. Born 1806, died 1859.'

Now take a moment to follow the direction of his unseeing gaze upriver. Firstly you will see the wide arches of Waterloo Bridge, described once as 'the most distinguished reinforced concrete bridge built in England'. Completed in 1945, this newcomer blocks IKB's view, but look through its arches and you will find the remains of the Brunel-designed bridge. Now surely a Brunel bridge smack in the middle of London – overlooked by the Houses of Parliament and more recently the London Eye – has got to be something to celebrate,

something to write home about? And yet Brunel's most celebrated biographer, L.T.C. Rolt, devoted less than one paragraph to its existence.

The commission to design a suspension bridge over the Thames came when Brunel was twenty-nine years old and already preoccupied with a raft of other engineering projects. Sitting in his office on Boxing Day 1835, he made a long list of these works including several railways, two docks and, of course, the Clifton Bridge, which he described as 'My first child, my darling'. Contrast that with his entry for this London bridge: 'Suspension Bridge across Thames – I have condescended to be engineer to this – but I shan't give myself much trouble about it. If done, however, it all adds to my stock of irons.'

Clearly IKB had poured his heart into the Clifton Suspension Bridge, and in return it had launched him on his meteoric rise as the country's pre-eminent engineer. So when the London job came along, he had already explored and solved all the design problems associated with a wide-spanned suspension bridge. Here was a man who relished original challenges, but one who didn't take much interest in revisiting old territory. I can see him now sucking on a cigar. 'Done suspension bridges … Next!' Which is a great shame because the surviving photographs and drawings of the Hungerford Bridge, as it was called, reveal a design that is as elegant as any you could wish for. Besides, for the hunter of Brunel's works it is all the more exciting to know that something of it has survived! Not only that, but there is still a twist or two in the telling of its story.

To begin at the beginning: the name Hungerford has nothing to do with the town in Berkshire. Instead it derived its name from Sir Edward Hungerford, whose house on the north side of the Thames was destroyed in a fire. The site then became the Hungerford Market and the bridge was constructed in the early 1840s, opening in 1845 as a pedestrian crossing from the southern side of the river. At 1,462 feet long it was one of the longest suspension bridges in existence. It had two side spans, each 343 feet long, and the centre one, at 676 feet, was actually longer than that required for the Clifton Bridge. There is a wonderful old grainy photograph of Hungerford Bridge, said to have been taken by the pioneering photographer William Henry Fox Talbot. Beyond a cluster of Thames barges, it reveals the elegant lines of the bridge's long cast-iron chains curving up to Italianate towers sitting atop solid-looking piers either side of the river. And between these towers, the walkway arches gently over the centre of the river. The artist James Abbott McNeill Whistler painted a similar image, 'Old London Bridge', and this was later issued as an etching in a series of sixteen widely referred to as the 'Thames Set'. So clearly the Brunel bridge was admired in its day.

A few historians suggest that the footbridge was well used by Londoners and that it was a great financial success. As well as charging the pedestrians a toll, money was also raised from the steamboat companies that used the piers and footings as landing stages. Certainly the opening of Waterloo station in 1848 must have significantly increased the foot traffic. But most sources concur that it was a commercial flop, partly because of the 'Great Stink' emanating from the river at that time. So what did cause its premature demise? Ironically, it was the Industrial Revolution itself. The Hungerford Market site was sold to the South Eastern Railway Co. for the construction of their London terminus and, in 1859, the year of Brunel's death, the railway company purchased the footbridge in order to extend the line from London Bridge over the river to Charing Cross. Presumably Brunel must have known what was in store for his bridge, and he would probably have approved of this onward march of progress with a steam railway replacing pedestrians. But what would he have thought of the girder bridge that replaced it? Described as 'squat', 'ugly' and 'aesthetically notorious', the railway bridge designed by the South Eastern's engineer, John Hawkshaw (later Sir John), is devoid of any of the finesse of Brunel's original. It is brutal and functional, all iron girders like a giant Meccano construction set. However, although the distinctive Italianate towers were demolished,

William Henry Fox Talbot's photograph of the Hungerford Suspension Bridge.

the new bridge does rest its heavy load on Brunel's original red brick and stone piers.

As the railway company was required to maintain a foot crossing, walkways were cantilevered out from either side of the rail bridge and these continued as paid crossings until the toll was abolished in 1878. Shortly afterwards, the upstream walkway on the Westminster side was incorporated into the railway bridge so it could be widened. That left only the downstream one, which was less than adequate. Described as narrow and noisy, it became the haunt of nocturnal muggers. Talk of a replacement goes back to when the Festival of Britain was staged on the South Bank in 1951 and a temporary Bailey bridge was strung across the river to cope with the crowds. Finally, in the 1990s, an architectural competition was held to finalise a design and cash was allocated to the project by the Millennium Commission. Even then it wasn't plain sailing, as London Underground dramatically announced that any unrecorded and unexploded wartime bombs still at the bottom of the river could be triggered by the piling work, resulting in a catastrophic flooding of the Underground – in particular the Bakerloo Line, which passes under the river near the bridge. Back to the drawing board and, with further funding courtesy of London's newly elected mayor Ken Livingstone, a redesign by architect Alex Lifschutz produced what is now widely accepted as a stunning new footbridge and landmark for London.

There is some confusion as to what to call the new bridges, for in fact there are two flanking the railway bridge. Officially they are the Golden Jubilee Bridges, but sometimes they are referred to as the Millennium Bridges and more often Londoners persist in lumping them together with the railway as the Hungerford Bridge. You can approach them from several directions. But to get there from the Brunel statue at Temple Place, carry on past Cleopatra's stone needle – once bleached by the sun of a clear Egyptian sky and now stained by years of London grime and pocked by shrapnel from a bomb dropped in a daylight raid in the First World War – and under the Waterloo Bridge to the new bridges.

This is engineering with a delicate touch. Slender white suspension masts, not unlike ships' masts, are linked by steel stay rods to the broad 15-feet wide decks or walkways. Seven million people a year are expected to use the bridges and from the outset they have proved to be popular public spaces in their own right, especially the upstream side, which has opened up new views towards Westminster, much to the delight of the camera-clicking tourists. This side was completed first, in May 2002, before the existing walkway on the downstream side was dismantled, to be replaced with a matching footbridge in September that year. Plans for spurs leading direct to the Eye and the Festival Hall, and also for a transverse connecting bridge passing through the hollow structure of the Surrey pier, have been shelved for now. A shame, because that really would have provided an exciting close-up encounter with IKB's work.

From either footbridge you get an excellent view of the brick piers and you can see how the new bridges rest or link with their footings. The rounded abutments rising above the level of the railway tracks are an embellishment added by Hawkshaw. But everything below the railway line is pure Brunel. The southern or Surrey pier was restored during the construction of the new footbridges, and the northern one on the Middlesex side, which displays a few alarming cracks in the brickwork, is also being worked on. The position of the piers as seen nowadays appears slightly lopsided, with the Middlesex pier almost butting the riverbank in comparison with the Surrey side. This is because the Embankment was constructed as an extension of the northern riverbank to house sewage drains and new Underground tunnels, and consequently it has cut off some of the curving sweep of the river. Take a look at the nineteenth-century images of the original bridge and you will see that both piers once stood much further out into the river. For a superb aerial view of the Hungerford site, or of the whole of London for that matter, go to the South Bank and take the short walk past the assorted living statues and other street performers up to the London Eye. From its slow-moving pods you can take in the bridges and the railway line disgorging into the modern Charing Cross complex.

The Bristol Connection

Finally, there remains one chain of events to clear up concerning IKB and his bridges, one that came about due to a most unexpected source. Sir John Hawkshaw – he of the 'squat, ugly' rail bridge – also happened to be on the committee of engineers working to complete the Clifton Suspension Bridge as a memorial to Brunel after his death. However, they had no chains. When work had stalled at Clifton in the 1840s, Brunel himself had recycled those chains for his Royal Albert Bridge over the Tamar at Saltash. So Hawkshaw purchased the now second-hand set of Hungerford chains for £5,000 and took them to Bristol. Which is where they are now, draped across the Avon Gorge.

EMBANKMENT STATUE

LOCATION: Temple Place, Victoria Embankment.

GETTING THERE: Between Underground stations Temple (closed Sundays) and Embankment.

Baron Marochetti's statue of a bare-headed Brunel was erected in 1877.

HUNGERFORD BRIDGE

LOCATION: Between Victoria Embankment and the South Bank.

GETTING THERE: Approached from either side of the river. Embankment is nearest tube station on the north, Waterloo on south.

Unlike Brunel's 1845 footbridge there is no fee for pedestrians using the Golden Jubilee Bridges, so enjoy the close-up view of the original red-brick and stone piers which now support the railway bridge. On the South Bank it is a short walk to the London Eye which has spectacular views of the bridges and London. Waiting times vary, with longer queues at weekends. (www.londoneye.com)

WESTMINSTER ABBEY

LOCATION: Westminster

GETTING THERE: Westminster or St James's Park Underground.

OPENING TIMES: Monday–Friday opening times vary depending on services so check website. Open Sunday for services only.

The 1868 memorial window to Brunel in the nave, enter via the west entrance and the window faces the deanery. A tourist trap, so get there early.

INFORMATION: 020 7222 5152 / www.westminster-abbey.org

3

PADDINGTON STATION

West London

Paddington station is one of Brunel's most spectacular achievements. But as an introduction to his Great Western Railway, putting Paddington first is a bit like putting the cart before the horse; for this glorious structure of iron and glass belongs to the latter period of Brunel's career, where it sits like an exclamation mark at the end of the line.

As well as being a great engineer, IKB was a very capable architect; a fact demonstrated throughout the length and breadth of the GWR. The wide wooden span of the Temple Meads terminus in Bristol, for example, was upheld as a wonder of its time, while the numerous smaller stations display great attention to detail. To some extent his taste in architectural styling was coloured by the fashions of the time and it could be argued that it veered towards historical pastiche, with more than a nod to the Tudors as epitomised by his signature skewed chimneys, or equally to the Italianate, or the rock follies of the numerous tunnel portals. Fortunately these throwbacks sit very comfortably within the landscape and they have weathered well, as the images within this book testify. But a London terminus had to be something special. Its design called for drama and presence, just as the Poet Laureate, the late Sir John Betjeman – a man who knew and loved railways – once said:

If the station houses are the equivalent of parish churches, then the termini are the cathedrals of the Railway Age. Most

companies, even if their origins were in provincial towns, were determined to make a big splash when they reached the capital.

At Paddington, Brunel was to take his railway architecture to another level. Unlike any of his previous structures, its graceful vaults of iron and glass appeal to the modern eye, successfully combining aesthetics and function without an ounce of superfluous engineering. So how did Paddington come to be so different from his other stations? One clue lies in the station's correct name, for this is in fact Paddington 'New' Station.

When the first GWR trains had departed from London back in June 1838, they had done so from a station situated to the west of Bishop's Road, just beyond the curving track which leads into the present station. The station buildings were a collection of unimposing wooden structures, designed by IKB but intended only as a temporary arrangement. In fact, the original station had been built in some haste following the collapse of talks between the GWR and the London & Birmingham Railway to link together at Kensal Green for a shared run into Euston. Undoubtedly it did not live up to Brunel's vision of the gateway to a modern transport system that stretched westwards all the way to New York. So when the GWR's directors gave the go-ahead for a more permanent replacement, his delight was self-evident, as is clear in this letter to architect Matthew Digby Wyatt, written in January 1851:

Paddington – a fitting terminus to Brunel's Great Western Railway.

The magnificent curves of the iron and glass roof at Paddington.

I am going to design, in a great hurry, and I believe to build, a Station after my own fancy; that is, with engineering roofs, etc. etc. It is at Paddington, in a cutting, and admitting of no exterior, all interior and all roofed in.

A large plot of land was purchased to the east of Bishop's Road, within the area defined on two sides by Eastbourne Street and Praed Street, on the edge of the city. This site did not line up with the existing track and the railway takes a fairly sharp turn to the right as it enters Paddington. More importantly, as it is situated within a cutting, it demanded an entirely fresh approach to station design. There would be no grand facade to trumpet the virtues of the GWR and instead Brunel devised an immense roof of iron and glass comprising three great arches or transepts.

Brunel's Greenhouse

In seeking the source of inspiration behind this design, many historians have pointed to Joseph Paxton's famous Crystal Palace. However, the dates simply don't fit. Author Adrian Vaughan flipped the argument about face when he asserted that it was actually the other way around:

> During 1848 Isambard designed a permanent station for Paddington, the construction of which began in 1849 … The design was unprecedented and set the style of industrial/railway architecture for many years to come. It predated Paxton's design for the iron and glass 'palace' for the Great Exhibition by at least eighteen months. Without any doubt Isambard's methods at Paddington were seminal for Paxton's design.

But Vaughan is claiming too much for his hero and neglects to mention that Paxton's first large iron and glass structure had been the Great Conservatory, built between 1837 and 1840 for the 6th Duke of Devonshire at his Chatsworth estate in Derbyshire. Known as the 'great stove' because it was so expensive to heat, this was the largest greenhouse in the world and through its design and construction Paxton developed the techniques that served him so well later on. There can be absolutely no doubt that Brunel, a man who kept a keen eye on all areas of engineering development, was familiar with the Great Conservatory a full eight years before work on Paddington began. It was Chatsworth that inspired him to build with iron and glass, and it was Chatsworth that led on to the Crystal Palace.

The links continue. Brunel and Paxton served together on the Building Committee for the Great Exhibition, which had been proposed by Prince Albert in 1849 and was held in 1851, and evidently they enjoyed a close working relationship. The design of a building to house the exhibition had been left open to competition and the committee members sifted through 245 submissions. Having rejected each and every one, the committee came up with its own collective 'camel' of a design featuring a series of low brick-built halls surmounted by Brunel's contribution, a vast central dome of iron and glass ballooning 150 feet above the rest. It looked absurd. It was then that Paxton produced his masterpiece, affectionately christened by *Punch* as the 'Crystal Palace'. Easy to erect, by virtue of its modular prefabricated construction, it was an extremely practical exhibition space with the added benefit of making a minimal impact on the Hyde Park site to the extent that several mature trees in the park were contained within its lofty central transept. Visitors to the 1851 exhibition were stunned by its scale and lightness of construction, and among them, IKB remained an enthusiastic supporter. He drew heavily upon Paxton's expertise in building with iron and glass to create what Betjeman described as 'Brunel's greenhouse' at Paddington.

Interestingly, the threads of Brunel's connection with the Crystal Palace continued after the end of the Great Exhibition. The newly knighted Sir Joseph Paxton recruited his help in the task of re-erecting the building in an enlarged form at Sydenham, in south London, where it was reopened by Queen Victoria in 1854. The new location lent itself to the development of elaborate ornamental gardens with many fountains and to supply them and the heating boilers with water, Paxton envisioned a pair of water towers, one at either end of the building, which also concealed chimneys leading from the boilers. Initially Paxton's assistant worked on their design but when Paxton brought IKB in to advise on the job he soon took charge, and under Brunel's guidance, two towers of cast iron were completed by the summer of 1856, each one supporting 1,500 tons of water.

The Crystal Palace remained a popular attraction for a further eighty years until the night of 30 November 1936, when a great fire ripped through the building. Brunel's towers survived intact because they stood apart from the main structure, but they were demolished at the beginning of the Second World War for fear that they might serve as navigational markers to the Luftwaffe. One was dynamited and the other dismantled. Only the base of the south tower and some of the water piping remains behind the Crystal Palace Museum. Alas, Paxton's

Brunel's water towers on either side of the enlarged Crystal Palace at its new site in Sydenham, south London. The towers provided water for heating and the fountains; they also concealed the boiler's chimneys.

The towers survived the terrible fire of 30 November 1936, but fell victim to fears that they might serve as navigation aids to the Luftwaffe bombers in the Second World War.

SOUVENIR OF THE CRYSTAL PALACE, DESTROYED BY FIRE NOV. 30TH 1936.   G.5900.

Great Conservatory at Chatsworth fared no better, having been torn down in 1923 because of the high heating and maintenance costs. At least Paddington station is still around, and in many respects it has survived remarkably unchanged in over 150 years.

Iron and Glass

There are various entrances into the station depending on whether you arrive by vehicle, on foot or via the Tube, but historically the principal entry point has been the Arrivals Road leading down from Eastbourne Street. Passing in through the Clock Arch, travellers are greeted by the figure of IKB keeping a watchful eye on his station. This seated statue, the work of sculptor John Doubleday, was commissioned by the Bristol & West Building Society and presented to British Rail in 1982 (Doubleday also produced a standing figure of Brunel for the other end of the line at Bristol). The statue once sat at the entrance on the concourse area, but was moved during the recent refurbishment. The Clock Arch leads straight on to Platform 1, resplendent in its new floor of French limestone, but the immediate impact upon the traveller is that great expanse of roof. That and the noise!

The roof of Brunel's train shed, to give it its proper name, comprising three transepts 700 feet long. The central span is the biggest at 102 feet 6 inches wide, and it is flanked by side spans of approximately 70 feet. Supporting the roof are wrought-iron beam arches spaced 10 feet apart, like the ribs of an upturned ship's hull. Originally they perched on slender cast-iron columns every 30 feet, although these were beefed up with steel in the 1920s to their present hexagonal girth. It is interesting to note that between the columns are two 'floating' arches supported by cross-bracing and not directly by the columns themselves. At the widening base of each arch is a simple 'petal' motif, which is both decorative and provides strengthening, and further up, the beams are pierced by clusters by the 'planets and stars' motifs to increase the sense of lightness. While the overall design for Paddington was IKB's, he was snowed under with other work and consequently he recruited the architect Matthew Digby Wyatt to carry out much of the detailed work and decorative treatment. (They previously collaborated on Swindon's Railway Village – *see* Chapter 5.)

The three long spans are punctuated by two crossways transepts, each one 50 feet wide, at one-third intervals lengthwise. These intersections interrupt the parallel ribs of the roof, creating an altogether more intricate geometry with their sweeping crossover girders. One of these transepts provides a clear line of sight to and from the Moorish windows and balcony of the directors' office overlooking Platform 1 – the 'Grand Departures' platform; grand perhaps because of the station's royal connections as the rail link between London and Windsor. To the left of the war memorial, the royal coat of arms can still be seen above the doorway leading to what was the Royal Waiting Room.

At the 'country' end of the station, a footbridge links the various platforms. The 'Arrivals' platform is situated on the far side and originally the space occupied by the central platforms was arranged as carriage sidings leading to turntables. It is from the footbridge, which has been much altered over the years, that you get the best overall view of the station and also a closer look at its roof. From the edge of the footbridge steps it is possible to tap some of the decorative bosses at the base of the flying arches, at which point you will discover that these are

actually in timber, not iron, another weight-saving device. In fabricating the roof, IKB brought in Fox & Henderson, the company that had constructed the Crystal Palace, and thus at Paddington we see an early example of a close collaboration between engineer, architect and builders. Above the footbridge are the glazed gable ends of the transepts. Decorated and strengthened by Digby Wyatt's surprisingly delicate and almost art nouveau tracery, they serve a dual purpose in letting in light and providing extra rigidity to the roof.

On a dull and dismal day, any comparison between Paddington and a cathedral seems to be stretching the point, but once the slanting fingers of the morning sun slice the air, the whole place stirs into life. The Victorian artist William Powell Frith set his great narrative painting 'The Railway Station' at Paddington, but he remained unimpressed by the aesthetics of the place and once commented, 'I don't think the station at Paddington can be called picturesque.' Behind his scene of

A seated statue of IKB by sculptor John Doubleday – its standing counterpart is at the other end of the line in Bristol.

The departures platform, swarming with straw-hatted boys heading to Eton in Edwardian times.

newly-weds, detectives making an arrest and an assortment of other travellers boarding a broad gauge train, the roof detail including the gas lighting is clearly shown and Frith's attention to detail provided a useful reference for the restoration of the roof in the 1990s, especially in the use of colour. In a weight-saving exercise, the corrugated iron cladding and glass has now been replaced with profiled metal sheeting and polycarbonate glazing.

It is easy for us to view Paddington from an historical and heritage perspective and to forget that this was state-of-the-art architecture. It was at places such as this that the Victorians experienced the 'shock of the new'. Nowadays Paddington is very much a 'comeback' station. Just as the stylists in the automotive industry have rediscovered the curve and toted it as the 'new' retro, so in the world of architecture there has been a move away from straight lines and blocks to a more anthropomorphic look. Paddington's elegant transepts create a light and economical structure that is at once both functional and aesthetically pleasing. By comparison, London's other great stations seem either pompously over-egged or over-engineered.

Reduction and Redevelopment

For a modern perspective on Paddington, I sought the opinion of Mark Middleton, a director of the architects Nicholas Grimshaw & Partners and the man who had overseen the refurbishment and redevelopment of Paddington for almost a decade:

For architects, its history plays a big part. Brunel is highly regarded of course as a man of conviction and the British like

that, and there is also the direct connection with the Crystal Palace. But it's also about the transfer of technology – the way the structure is braced by Eastbourne Terrace and how Brunel used his ideas on boat-building. That's what gets the juices flowing for architects.

He describes the main thrust of the recent developments at Paddington as the 'architecture of reduction'.

By removing much of the accumulated clutter, including the cumbersome information boards, his team have restored dramatic and unimpeded views of the roof. But it is away from the footbridge and at the other end of platforms, at the 'town' end, that the new work is most evident. Different parts of Paddington have their own names and the main public concourse area is still quaintly referred to as 'The Lawn'. According to the stories, this name came about because the stationmaster once had a garden on the spot, or possibly it was the site of former hotel gardens. Either way, it has been transformed by the £42 million Phase 1, which was completed in September 1999.

In a bold statement of modernity, a 150-feet-long glass screen now separates The Lawn from the hubbub of the station and has created a quieter, climate-controlled environment. Here you can sit in leatherette comfort sipping a mocha coffee, and a passably good one at that, sandwiched between the trendy Café Ritazza and a sushi bar. The glass walls block out some of the background noise, while all around people speak loudly into their mobile phones. 'No, no, I'm in London,' barks the man on the next table, 'I have the keys with me.' Other travellers have more respect for this cube of calm and they sit silently, nodding to the beat of a hidden iPod. Escalators lead up to mezzanines for direct access into the Hilton London Paddington, while on the ground floor, passengers check in to the Heathrow Express in an ambience more befitting an airport. The result, I have to say, is a triumph; a modern space that sits comfortably within the concept of Brunel's station without impeding the views or overpowering it in any way.

Not mentioned so far is Paddington's fourth span. At the time of the Brunel bicentennial in 2006, there was a very real danger that the fourth span was about to be demolished. Running parallel to IKB's original three, the fourth had been added on the north side in 1916, along with improved below-platform facilities for storage and new stabling for an army of 500-plus working horses, which pulled the delivery vans. Built of steel, it echoes IKB's design, but only loosely. From one end, a gentle slope leads through an archway emblazoned with the old GWR crest and up to Praed Street. The Phase 2 redevelopment of Paddington had entailed the loss of Span 4 in order to make way for new 'transport interchange facilities'. In plain speak, it was to provide additional platform space, in particular for the new Crossrail cross-London line, as well as for the realignment of platforms 9 to 14, an enlarged concourse area including a canal-side entrance, new access for taxis from Bishop's Bridge Road and new commercial space above. When plans were first submitted in 2000, the artist impressions revealed a lightweight structure with transparent walls. But the loss of Span 4 stirred up the industrial heritage groups, who soon found themselves at loggerheads with the more purist Victorian Society. At that time revised proposals had been approved by the planning authorities and, with English Heritage granting Listed Building Consent for demolition, the fate of Span 4 appeared to be sealed. Only the timing seemed uncertain, as Mark Middleton revealed:

> It is closely linked with the Crossrail requirements. They need track pathways and we have to demolish Span 4 in order to achieve that. We could be starting quite soon, maybe in 2007, but as we can only close one or two platforms at a time it may take three or four years to complete.

For much of the twentieth century the archway into Span 4, with its coat of arms, welcomed millions of travellers on their journey westwards. Its demolition would have been a sad loss. Phase 2 was to have been a massive step in the history of Paddington station, and one that would have drastically altered its character. However, the good news is that Span 4 was reprieved and a new site for Crossrail is being created by excavating beneath Eastbourne Terrace on the western side of the station. Work was still in progress at the time of going to print, and the new Crossrail station is due to open in 2017.

Around Paddington

Although Paddington is described as a station that is all inside with no recognisable outside, it is surrounded by a variety of interesting buildings. So sneak past the Paddington Bear statue and the souvenir sellers – and yes, some of the tourists really do think the station was named after him – and head outside for a closer look. Rising above what

is currently the 'Departures Road' at the end of Span 4 is the wonderful 1932 art deco facade of the GWR's offices, originally known as the 'Arrivals Side Offices' and, since refurbishment in 1987, as Tournament House. Designed by P.E. Culverhouse, architect for the railway company, this steel-framed building is a remarkably modern design distinguished by its GWR roundel or 'button-hole' logo, the subtle diamond motifs and a row of lights housed within dark upturned shells.

Beyond and into Praed Street itself there is the imposing Great Western Royal Hotel, the nearest thing Paddington station has to a frontage. Opened in June 1854, it was designed by Philip Hardwick in the style of a French chateau with tall corner towers, albeit with classical pediment and frieze. The hotel was managed separately from the railway, although the first chairman of the board was none other than IKB, and it was very much part of his vision of an integrated transportation system. Once one of the most sumptuous hotels in England, it was redecorated in the art deco style in the 1930s, and for the hotel's 125th anniversary in 1979, the new Brunel Restaurant was added. The most recent refurbishment of this Grade II-listed building was completed in 2002.

On the opposite side of the road is Sir John Fowler's 1920s facade of the Metropolitan Railway station. The station itself dates back to 1860, when it was the western terminus on London's first Underground line. It is now part of the District and Circle Line.

Paddington's Iron Bridge

This journey through the work of IKB has thrown up several surprises along the way and Paddington proved to be no exception, thanks to the chance discovery of a hidden bridge behind the station. In 2004, Dr Steven Brindle of English Heritage was examining Brunel's notebooks while researching a new book on Paddington, when his attention was caught by designs and records for load testing for the cast-iron beams of a Paddington canal bridge dating from 1838. Letters concerning the bridge, written by IKB to the Grand Junction Canal Co., were also found, but no one knew if the structure was still in existence. By an extraordinary stroke of luck, it was located beneath the brickwork of the modern Bishop's Road Bridge, which was earmarked for replacement as part of improvements. The railings and upper structure had gone, but incredibly the iron arches were still intact. Steven Brindle told me that it is now in storage ready for restoration and relocation.

PADDINGTON STATION

LOCATION: Praed Street, W2.

GETTING THERE: Various underground lines.

When you have finished gawping at the magnificent glass roof, you can find the seated figure of Brunel at the old taxi drop-off entrance leading to Platform 1. From there, go over the footbridge to other platforms, or stay with Platform 1 for a coffee beside the plaque to IKB (or, if you prefer, a little sushi on The Lawn). Spare some time to take in the wonderful 1932 art deco facade of the GWR offices on the approach slope to Praed Street. Around the corner is the Great Western Royal Hotel, now the Hilton London Paddington, and the 1920s facade of Paddington Metropolitan station. For the rediscovered iron bridge site, from Praed Street follow round the east side of the station to the Grand Union Canal. Nothing to see of it now, although there are plans to re-site it as a footbridge.

INFORMATION: www.networkrail.co.uk

THE CRYSTAL PALACE MUSEUM

LOCATION: Crystal Palace Park, Sydenham, SE19 2BA.

GETTING THERE: Crystal Palace station. If travelling by road, it is on Anerley Hill off the A212.

OPENING TIMES: Summer Saturday/ Sunday 11.00–16.00, winter 11.00–15.00.

Remains of the base of the south tower lie behind the museum.

INFORMATION: www.crystalpalacemuseum.org.uk

4

PADDINGTON TO SWINDON

Brunel's Billiard Table

In the 1830s, the race to link the towns and cities of Britain with a tracery of glittering lines of iron was the result of two technological breakthroughs: the introduction of new methods of smelting iron, and the development of efficient steam engines. The coming together of these two factors made the railways practicable and, crucially, affordable. As John Cooke Bourne wrote in the introduction to his 1846 book *The History and Description of the Great Western Railway*:

> Great inventions, like great discoveries, are quite as often due to the general turn of mind of the age in which they are produced … Now it is probable that there has never been a period in the world's history, before the present in which the locomotive engine could have been invented, or, having been invented, could have been generally used.

Fortunately for IKB, he happened to be in the right place at the right time. When the first public railway – George and Robert Stephenson's Stockton & Darlington line – opened in 1825, Brunel was eighteen years old, and when he was appointed as engineer to a new railway between London and Bristol in 1833, he was still only twenty-six. The whirlwind speed of the railway's growth was unprecedented and while the construction of the canals had occupied half a century, the

principal railways of England were built in less than a decade. In the 1836 parliamentary session alone, fifty-seven Bills were presented for the construction of more than 1,500 miles of railway, forming one-sixth of the private business of the House. In an instant the railway system outstripped every speed that had ever been heard of before, it brought towns and cities closer together and transformed a population of predominately country dwellers into townies.

The plan for a Bristol Railway was first proposed in 1832 and IKB submitted a tender to conduct the survey work in February the following year. In August 1833 he was confirmed as engineer and for the first time he wrote in his journal the title 'Great Western Railway'. From the start Brunel was adamant that this railway would be unlike any other. It had to be the best, not a carbon copy of the existing northern railways. His passengers were of a different calibre altogether. They lived in fine cities such as Cheltenham and Bath; they were 'persons living upon their incomes', and he would ensure that they could take their coffee whilst travelling at 45mph. With so few precedents to follow, he had, in effect, a blank sheet on which to plan his great work, and with the directors in the palm of his hand, he was about to take them for the ride of their lifetime. 'The route I will survey will not be the cheapest – but it will be the best,' he asserted. Not for nothing did the GWR become known as God's Wonderful Railway and the line from London to Bristol has survived as IKB's masterpiece and most enduring monument – 118 miles of the finest track in the world which, barely altered in over 150 years, was the first to carry modern high-speed trains at 125mph.

Deciding the Route

There were two options; south of the Wiltshire Downs or north and around the top of them. The route to the north offered the greater opportunity for trunk lines to destinations such as Oxford, Cirencester, Cheltenham and, via Gloucester, to Wales. It was also the more level way. The impression when travelling on the railway nowadays is of a straight line going from Bristol to London, but when seen on a map, the route curves upwards within ten miles of Oxford. This caused some wits to refer to it as the 'Great Way Round', but those in the know preferred the accolade 'Brunel's Billiard Table', as the incline between London and Didcot was only 1 in 1,320.

IKB undertook the surveying personally, working to a punishing schedule to cover the ground, often riding for up to forty miles a day. Later he had a special horse-drawn coach built, complete with a drawing board and bed, and this became known by the railway workers as the 'Flying Hearse'. This set the pattern for the rest of his life – living on the job and working incredibly long hours. Surveying for a railway was no easy matter, as many of the landowners were dead against this new-fangled contraption blundering across their land, with some hiring watchmen to keep the railway men away. There were even cases of mantraps being set against them. IKB was also careful not to attract too much attention to his survey work in case land prices along his prospective route went through the roof.

Brunel had already experienced rail travel on the Liverpool & Manchester Railway, and he wasn't impressed. The Stephensons had adapted the 4-feet 8-inch gauge already used by many northern collieries but the result, in his opinion, was a decidedly bumpy and unsatisfactory ride. He felt that the key to a smoother ride lay in a wider gauge and a far more rigid track with the rails held firmly in place on wooden bearers attached to heavy piles driven into the ground. As he wrote to his directors:

> The resistance from friction is diminished as the proportion of the diameter of the wheel to the axle is increased … we have therefore the means of materially diminishing the resistance … by simply widening the rails, so the body of the carriage might be kept entirely within the wheels, the centre of gravity might be considerably lowered and at the same time the diameter of the wheel be unlimited. I should propose 6 feet 10 inches or 7 feet as the width between the wheels.

By chance, or possibly by design, his submissions for the second reading of the GWR Bill omitted any clause specifying a gauge. He settled on 7 feet (plus an odd quarter-inch for leeway). Was this decision just bloody-minded or was he blind to future problems of incompatibility with other railways? I believe that the young Brunel was so self-assured, so cocksure of the superiority of his system, that he firmly believed all the other railway companies would come around to his way of thinking.

The Bill for the construction of the Great Western Railway first went before Parliament in early 1834, but was thrown out by the House of Lords, partly because of objections to the proposed site for the London terminus at Vauxhall. The following year the company went back with alternative proposals, including the idea of sharing access into London with the London & Birmingham Railway and a joint terminus at Euston (later abandoned in favour of Paddington). For forty days it was examined by parliamentary committee and IKB became the star attraction as he skilfully and calmly responded to their probing questioning. Royal Assent for the GWR was eventually granted on 31 August 1835 by William IV. There were two boards of directors, one for Bristol and one for London, and the plan was to construct the line from either end. However, as the Bristol end had to contend with the more difficult terrain, the railway spread westwards from London much faster and as each new station was added, it became the temporary western terminus for the railway. To simplify matters, I have treated the GWR as a continuous east-to-west journey.

West London: Wharncliffe Viaduct

Almost immediately after leaving Paddington, the line heads westward and seven miles or so later, just beyond the station at Hanwell, IKB was faced with the wide Brent Valley. Here he constructed the imposing Wharncliffe Viaduct, which at 896 feet long and 65 feet high is the largest brick-built structure on the GWR. It consists of eight semi-elliptical arches of 70-foot spans and mounted on pairs of elegantly tapered red-brick pillars or piers capped with sandstone cornices, evoking Brunel's favoured Egyptian style also seen on the towers on the Clifton bridge (incidentally, as with Clifton's abutments, these brick constructions are hollow). Apart from being decorative, the cappings on the piers also served as supports for the wooden formers or centrings, which held the arches until the mortar had set. Below ground, the foundations for the piers descended to the strong blue London clay, with the base of each pier having a footprint area of 252 square feet. The viaduct is named after Lord Wharncliffe, the chairman of the House of Lords Committee who had supported the building of the railway, and his coat of arms can be seen high on the southern face of the brickwork.

Bourne's engraving of the viaduct depicts a self-assured structure brazenly striding through an otherwise rural landscape complete with cowherds and wagons. It is a bold architectural statement and a portent of things to come. Today the urban sprawl of Ealing and West London is snapping at the heels of the great viaduct, although the immediate area is a public park. The major difference since Brunel's time is the addition of a third pier on the northern side, added in the 1870s when the original double broad gauge track was extended into four standard lines.

A short distance west of the viaduct, the railway line passes over the Uxbridge Road at an awkward intersection with the Guildford to Brentford road, where we encounter one of the few instances in which IKB's work did not stand the test of time. His wide skew bridge consisted of stone abutments with cast-iron girders supported on two rows of eight masonry columns, but unfortunately this unusual bridge was damaged in 1847 when the timbers below the track caught fire.

J.C. Bourne's depiction of the imposing Wharncliffe viaduct, which strides across the Brent Valley on elegantly tapered redbrick piers; it is the first major work on the railway heading westwards from Paddington.

The Wharncliff viaduct as it is today, with the coat of arms of Lord Wharncliffe, who had supported the GWR Bill in Parliament, emblazoned on the brickwork.

Continuing due west, the line arrives at Slough, a significant station that also served the population of Windsor (with its obvious royal connection) and nearby Eton School. (Brunel's wrought-iron bowstring bridge crosses the Thames at Windsor – *see* guide for location.) Until the building of a direct spur line to Windsor, Queen Victoria would alight at Slough and accordingly the station featured 'two reception rooms, very handsomely furnished' for HRH's use. Slough also boasted the magnificent Royal Hotel, which catered for travellers described by Bourne as being of 'very high description'. Finished in the Florentine style, IKB may have had a hand in its design.

Maidenhead Bridge

After leaving Slough, and twenty-two miles out of Paddington, the railway reaches Taplow station on the eastern side of Maidenhead (the new Maidenhead station being constructed about thirty years later). It was this section of the GWR from London that was the first to open for service, in June 1838, and it was to remain the westbound extent of the rails until July the following year. The delay in extending the line was caused by the 300-foot-wide Thames a quarter of a mile further on, and it was here that IKB built his celebrated Maidenhead Bridge, featuring two of the widest and flattest arches ever carried out in brick. With a central pier standing on a shoal midstream, the two main semi-elliptical arches each have a span of 128 feet with a rise of just 24 feet 6 inches, while four smaller semi-circular arches at either end join with the approach embankments. The result has been described as 'visually the most pleasing and technically the most daring of Brunel's designs in brickwork'. But at the time of its construction, it was greeted with scepticism by some experts, who predicted that the brick arches would not actually stand at all, let alone bear the weight of trains. The 1935 GWR publication *Track Topics*, charmingly subtitled *A Book of Railway Engineering for Boys of All Ages*, takes up the story:

> There was doubtless joy in the hearts of those experts when, on the centrings of the bridge being removed, the eastern arch disclosed a slight distortion. It was actually a separation of about an inch in the lowest three courses of bricks, due to the centring being eased before the cement had time properly to set. The contractor admitted that he alone was to blame and the fault was duly remedied.

Once the remedial work had been carried out, IKB ordered that the centrings – wooden supports – should be eased but left in position over the winter. However, when a severe storm in the autumn blew them away, the bridge was left standing as straight and true as it does today. If you visit the bridge now, just to the south of the old road bridge and visible from it, you will see that the rail bridge was widened when the tracks were quadrupled in the 1890s. The scene couldn't be more picture postcard, with the pleasure craft on the river flanked by the manicured lawns, smart homes and summerhouses of the well heeled. This is a very busy stretch of the mainline and the fast trains appear out of the blue, with only a distant bark of horns announcing the flash of carriages that leave a vacuum of quiet in their wake.

Sonning Cutting

By July of 1839, the tracks had been extended as far as the village of Twyford, where a wooden station was constructed on the north side of the track, and beyond Twyford lay the wide hills at Sonning. Initially IKB had proposed taking the line slightly to the north and under the high ground by means of a tunnel to keep it out of sight of an influential landowner. But he decided on a deep cutting at Sonning instead: a

At the Maidenhead Bridge, the two main semi-elliptical arches each have a span of 128 feet with a rise of just 24 feet 6 inches.

two-mile swathe through the landscape varying in depth from 20 feet to nearly 60 feet. It was a massive feat of engineering, with an army of up to 1,200 navvies excavating 700,000 cubic yards of material, often in appalling weather. Peering down into the cutting from the A4 road bridge, the trains seem so tiny that it is hard to appreciate its scale. Next stop, Reading.

Reading's One-sided Station

The present station at Reading is no great shakes, but the original was one of two single-sided stations laid out by Brunel, the other being Slough. When the railway arrived in 1840, most of the town lay on the south side of the line and he devised an unusual one-sided layout for both 'up' and 'down' trains – in fact, two stations on the same side of the track. The advantage of such a layout is that the passengers did not have to cross any track, regardless of the direction they wished to travel. But set against this, a train had to pass over the lines going in the opposite direction and as trains became longer, they risked straddling

This rare image of the single-sided station at Reading shows how the lines had to cross over each other to reach the platforms.

Varying in depth from 20 feet to nearly 60 feet, Sonning cutting is so big that it is difficult to appreciate its scale. It is shown here as the last broad gauge train pauses for a photograph.

the lines. Just describing the track layout is complicated enough, but it should be remembered that when IKB came up with this configuration, there were no ground rules for getting passengers to their trains. It had to be worked out through trial and error and the one-sided layout fell into the latter category. Sure enough, the old station at Reading was replaced in 1896.

Basildon and Moulsford Bridges

Leaving Reading, the line soon enters the Thames Valley, where it closely follows the Thames, passing over it in two places via a pair of fine brickwork bridges at a skew. At Upper Basildon, the Thames bends to the west and the bridge crosses into Oxfordshire at an angle of 15 degrees, while two miles away, the Moulsford Bridge re-crosses the river at a sharper 45 degrees. Both feature broad elliptical barrel arches with face-rings, cornices and copings of creamy limestone. Of the two, the bridge at Moulsford is more accessible thanks to a footpath leading down to the river, and standing beneath the arches you can appreciate the complicated curving of the brickwork – quite fantastic. You can also see the gap where the line was quadrupled by building a parallel bridge alongside, and slightly separated from, the original.

On the other side of the main A329, on the western end of Moulsford Bridge and across the shallow cutting, is IKB's hotel for the old Wallingford Road station. This small station was superseded by the Cholsey and Moulsford stations in the 1890s. From here the line curves north-west towards Didcot (*see* Chapter 10) and roughly halfway between Paddington and Bristol, it reaches the village of Steventon. This minor station served as the terminus for a short time in 1840, but more importantly its central location was chosen as the weekly meeting point for the company's directors when the London and Bristol boards combined in 1842. The old station is gone, but the stationmaster's house

still stands in Station Road, where it looks out on to the busy railway. This attractive stone building is now a family home.

Onwards and westwards, the countryside leading into the Vale of the White Horse was described by Bourne as 'more open, less thickly populated, and in some respects less interesting'. The track passes slightly to the north of Wantage, and south of Farringdon and Shrivenham, before dipping to the south-west and along the vale where the ancient horse gallops across the escarpment. Into Wiltshire, the early trains stopped at Wootton Bassett Road three miles beyond the small market town of Swindon. Initially no station had been provided for Swindon, which was perched on the top of a small hill, and for a while its inhabitants were blissfully unaware of their role in Brunel's Great Western Railway.

The bridge at Moulsford was doubled in width by the addition of a parallel bridge alongside, hence the gap. (Ute Christopher)

WHARNCLIFFE VIADUCT

LOCATION: Hanwell, south of Ealing. (OS map 176:150804)

GETTING THERE: About 400 yards west of Hanwell station. By car leave M4 at either junction 2 or 3, heading north to join the A4020 Uxbridge Road. To the west of Hanwell, you will see it to the north of the main road. Park near the recreation ground.

Named after Lord Wharncliffe, this imposing viaduct spans the Brent Valley. Originally 30 feet wide to accommodate two lines of broad gauge, it was widened in the 1870s.

THAMES BRIDGE AT MAIDENHEAD

LOCATION: Visible to the south of the A4 road bridge over the Thames on the east side of the town. (OS map 175:902811)

GETTING THERE: Leaving Maidenhead, turn right immediately after the river. There is some parking on street, but nearer to the bridge it becomes a private road.

Brunel's wide and impressively flat brick arches at Maidenhead confounded his critics, who were convinced they would fall. You will find a plaque on the southern face of the bridge, on the east bank.

THAMES BRIDGE AT WINDSOR

LOCATION: North of Windsor, Berkshire. (OS map 175:961773)

GETTING THERE: A reasonably short walk from Windsor Royal station. By road, junction 6 of M4 and south on A332, turn left into Windsor and left again up to the swimming pool on Stovell Road and car park. Walk across the pleasure park.

Crossing the Thames on the Slough to Windsor railway spur, this is a surprisingly modern-looking wrought-iron 'bowstring' bridge. Originally the approaches were constructed of timber, later replaced by brick, while cast-iron cylinders supported the bridge itself.

MOULSFORD BRIDGE

LOCATION: North of Moulsford, Oxfordshire, between Cholsey and Goring.

GETTING THERE: A329 Wallingford Road. From the north, look for the sign 'Moulsford Preparatory School – 200 yards' on the left. A lay-by is beyond it on the right (just past turning for the school if coming from the south). Walk back to the sign and there is a smaller sign for the Thames Walk footpath going past farm buildings to the bridge.

A wonderful structure with its curving brickwork. Best appreciated from the west bank, but you can also take the footpath from South Stoke on the east side.

STEVENTON

LOCATION: Steventon, Oxfordshire. Four miles west of Didcot.

GETTING THERE: A34 from either M4 or Oxford, take the B4017 into Steventon and turn just north of railway and humpback bridge into Station Road.

Situated halfway between Bristol and Paddington, this small village was selected by the board of the GWR for their meetings in the Brunel-designed Station House. Station buildings are gone, but the handsome house still stands as a private residence.

SONNING CUTTING

LOCATION: Two miles north-east of Reading, between Sonning and Woodley Green. (OS map 175:752742)

GETTING THERE: A large roundabout on the A4 towards the western end of the cutting provides a good vantage point. It's a busy road and parking is awkward, so you may prefer to seek out one of the other crossing points.

Two road bridges cross the cutting at this point and the triple-arched westbound bridge at the roundabout is Brunel's. Up to 1,220 men toiled with shovel and pick to carve this 50-feet-deep scar through the Berkshire hillside.

5

RAILWAY TOWN

Swindon: Pulling Power!

Although not exactly halfway between Paddington and Bristol in terms of miles, Swindon is at the point where the undemanding gradients of 'Brunel's Billiard Table' give way to more challenging terrain in the west. Sitting like a spider at the hub of a web, Swindon became the driving force of the GWR and as a result its population grew from 2,000 inhabitants in 1800 to 45,000 by the end of the century. Even so, IKB's quest for pulling power got off to a bad start and he was faced with a power struggle in both the boardroom and on the tracks.

Brunel was a genius and arguably the most gifted civil engineer of his age – and he knew it! So supremely self-confident was the 'Little Giant' that he demanded absolute control of every aspect of every project, and he expected to be the best at everything he put his hand to. Such absolute faith in his own abilities sometimes blinded him to the fact that he was simply too busy to be the master of everything. His ideas concerning the design of locomotives produced disappointing results, and when this failing was combined with shareholders' concerns regarding track gauge and the smoothness of the ride, IKB found himself in the firing line. Literally.

The GWR had no in-house engine works at first, so IKB produced his own set of requirements for outside contractors to follow. These stipulated a speed of 30mph to be considered the 'standard velocity', while the rate of the piston must not exceed 280 feet per minute, and the weight of the engine must be under 10.5 tons if mounted on six wheels or 8 tons on four. By imposing these conditions he made it impossible for the engine builders to produce efficient locomotives, and he also failed to take advantage of the extra width of his broad gauge. This has been described as 'the greatest and most inexplicable blunder of his whole engineering career'. Fortunately help was at hand.

In July 1837 a young mechanical engineer named Daniel Gooch wrote to Brunel applying for a job and he was immediately appointed as the chief locomotive assistant. Gooch had been with Robert Stephenson's Newcastle locomotive company and had worked on two 5-feet 6-inch locos intended for the New Orleans Railway, an experience which converted him to the broader gauge. 'I was delighted', he later recorded, 'in having so much room to arrange the engine.' As these two locos were never delivered to the American customer, one of Gooch's first acts on behalf of the GWR was to acquire them and modify them for the 7-feet gauge. The first, *North Star*, was delivered by barge to Maidenhead to await the arrival of the rails. She was the first GWR locomotive not built to IKB's requirements and she eventually proved to be a real thoroughbred. As Gooch wrote, 'We have a splendid engine of Stephenson's, it would be a beautiful ornament in the most elegant drawing-room …' And with trouble brewing for IKB, *North Star* was to save his bacon.

In common with many of the railway companies, the GWR had a large contingent of northern shareholders; businessmen predisposed to invest in railways. This 'Liverpool Party' didn't much like Brunel or his broad gauge, or the idea of his *Great Western* steamship with its railway links stealing their thunder as the principal regional port. So they set out to make trouble wherever they could and began by putting pressure on the GWR directors to have Brunel's wings clipped by suggesting the appointment of a second engineer. First blood was drawn at an acrimonious seven-hour meeting in Bristol on 15 August 1838. This failed to come to any decisions and was adjourned until October – although the directors subsequently agreed to have an independent engineer report back on the railway. Robert Stephenson declined to undertake this work on the grounds of his close friendship with Brunel, and Nicholas Wood agreed in his place. He was later joined by John Hawkshaw of the Manchester & Leeds Railway and by IKB's greatest detractor, the Revd Dr Dionysius Lardner.

Hawkshaw's report appeared first and although critical, it lacked any useful or constructive guidance, leaving IKB more than enough scope to dismiss it. Next came Wood's long-winded report, which attacked IKB for his use of timber piles to create rigidity for the track as the rails tended to sag in-between, causing an uncomfortable see-saw motion. Brunel had already recognised this problem and had ordered their removal. However, this was not enough to halt the campaign of whispers and the appointment of a second engineer remained on the cards despite his threats to resign if this happened. Perhaps the most damaging evidence against him came from Dr Lardner's experiments with the company's flagship, *North Star*. Lardner asserted that the engine was capable of hauling 82 tons at 33mph, but only 16 tons at 41mph, and he recorded excessive fuel consumption at higher speeds, which he attributed to the increased wind resistance of the wider broad gauge engines. Of course Brunel and Gooch knew this to be nonsense, although they were aware of the fuel situation. They conducted their own experiments with *North Star* and it didn't take them long to discover that the source of the problem lay with the blast pipe, which was too small. Once modifications were made, they unleashed *North Star*'s full power and on 29 December a group of directors was whisked from Paddington to Maidenhead at an average speed of 38mph. At the next meeting, in January 1839, Lardner's evidence was triumphantly dismissed by IKB, who was left to get on with his job as sole engineer of the GWR.

Clearly the company needed its own engineering works to build and service the engines and both Reading and Didcot were suggested, although Gooch preferred Swindon. It was, he argued, the most obvious place in which to change locomotives, with the single driving wheel locos running at high speed on the level track from Paddington replaced by ones with smaller driving wheels to tackle the ups and downs westwards to Bristol. After a site visit he wrote in his diary: 'Mr Brunel and I went to look at the ground, then only green fields, and he agreed with me as to its being the best place.' It wasn't long before those green fields gave way to the Swindon Railway Works, or the 'Works' as it became known in the town. One of the earliest eyewitness descriptions comes from J.C. Bourne:

> At some distance west of the passenger station, on the south side of the line, is the Engine Depot; its arrangements are upon a large scale, and capable of accommodating about a hundred engines. These consist of the engines in actual use, of the stock of spare engines, and of those undergoing repair. At this station every train changes its engine, so that from this circumstance alone, at least twice as many engines are kept here as at any part of the line.

Brunel's engine house at Swindon was demolished in the 1920s to make way for more modern workshops.

His illustration shows Brunel's engine shed as a rectangular wooden building, 400 feet long and 72 feet wide, with both ends open. On four lines of rails it held as many as forty-eight locos with their tenders, all ready for business. Louvres in the roof allowed the steam to escape. At right angles to this shed was the engine house, 290 feet by 140 feet, where the engines received light repairs, while major repairs and engine construction took place in the erecting house.

The Works

The Swindon Works grew quickly. In 1843 the workforce totalled 423 men, and five years later the workspace had doubled and the workforce increased to 1,800. The first of Gooch's highly successful home-grown locomotives began to emerge with *Fireball*, *Meridian* and *Spitfire*, and the 2-2-0 *Firefly* evolved in a slightly enlarged form into the 2-2-2 Firefly class. In 1846 the first of the Iron Duke class was constructed in just thirteen weeks, and with massive 9-foot driving wheels, it attained a top speed of 60mph. Other activities were also consolidated on the site and from 1868 the Carriage and Wagon Works accounted for all GWR carriage construction, drawing upon a wide diversity of skills, from traditional coach-building to upholstery. At its peak in the early twentieth century, the works occupied 320 acres of land, including 77 acres of covered sheds, and employed a workforce of 12,000. IKB's original wooden structures had been demolished by 1929 to make way for newer workshops.

Working at the railway works has been described as being in 'Hell's kitchen' but, as Felicity Jones, the curator at the Steam museum explained, it depended very much on which department you worked in: 'Life in the foundry was quite horrendous and very different from the offices, for example. But no doubt when people first entered the works they must have found it extremely dirty and noisy.' Just imagine being inside a boiler while the riveters hammered away from the outside:

> Some areas of the works could be dangerous, especially in the foundry or the shunting yards. In this respect it was no different to any other heavy industry at that time and, without the benefits of modern health and safety protection, accidents sometimes happened. People were hurt moving heavy equipment, even whole trains. But the GWR always looked after its people.

Lose a limb and the carpentry department would knock up an artificial replacement. Even so, there was one area of industrial hazard that couldn't be easily mended. Asbestos was used for the cladding of boilers in the loco department and later as lagging on diesel multiple units: 'There are horrific stories of the white powder dust being thrown about by the men in snowball fights. People were simply not aware of the hazard then.'

In addition to the men, the works took in a number of women, most notably in the offices, laundry and upholstery departments, and in much greater numbers during both world wars. During the Second World War, Swindon was called upon to manufacture a wide range of equipment including tanks, road vehicles, landing craft, munitions of all sorts and even secret miniature submarines built for the Admiralty – all this in addition to the vital work of keeping the trains running. The post-war nationalisation of 1948 saw the GWR name replaced by British Rail, Western Region and a major reorganisation of the works came in the 1960s with the end of steam and the arrival of diesel traction. The workforce was reduced and the Works saw a gradual decline until the mid-1980s, when the closure of British Rail Engineering Ltd was announced, dramatically coinciding with the GWR 150th celebrations. It was a bitter blow to the town that owed so much to its association with the railway.

In recent years, the sprawling Works site has been transformed and as you walk around, it is hard to imagine the hubbub of former times as many of the old buildings have gone. A few have survived as part of a shopping centre – a 'retail outlet' in modern parlance – or as offices. Fortunately there is also Steam, the Museum of the Great Western Railway, which opened in 2000. Carry on through the various workshop areas and you head past, or even under, the 120 tons of the mighty 1923 *Caerphilly Castle*, to reach the broad gauge exhibition space. Here you come face to face with a figure of the great man himself, unerringly lifelike in the familiar 'chains' pose and much photographed by the visitors. Behind him is the imposing *North Star*, sadly not the real McCoy, as only the 7-feet diameter driving wheels survived from the original, which was scrapped in 1906. This replica was built for the 100th anniversary of the Stockton & Darlington line held in 1925, and ten years later it was wheeled out to feature widely in the GWR's own centenary and starred in the company's 1935 film *Romance of the Railways*. More than any other exhibit this great engine, clad in polished wood and capped with brass fittings and a tall chimney rising into the rafters to echo IKB's hat, conveys the might of the broad gauge and is an indication of what might have been. And take a moment to look at the fireman and driver's position; the footplate, completely exposed to the soot and smoke from the engine, not to mention all that the weather had to throw at them. The Brunel hunter will find some finely detailed models of broad gauge locos including *Emperor* and *Iron Duke*, as well as a fascinating cabinet containing what looks like a selection of cheeses in a delicatessen. In fact this is IKB's collection of stone samples found in districts served by the GWR, his swatch-book of materials if you like, a personal resource on the building blocks of the fabric of the railway. The remainder of Steam goes on to tell the story of the works and of the GWR in particular, but I have to say that for me there is one thing lacking in this almost sterile atmosphere – that distinctive and evocative smell of hot oil, smoke and live steam.

Swindon Railway Village

Leaving the museum and walking under the main railway line through a long and gloomy pedestrian tunnel – which has most definitely retained all its authentic odours – you reach the Railway Village. This is also the way through to Swindon station along Station Road, predictably. The station was opened in 1842 and, as a busy junction with services to

Gloucester, it was later renamed as Swindon Junction to avoid confusion with the Midland & South Western Junction Railway's station up the road at Old Town. The original station designed by IKB featured brick and stone buildings either side of the track, linked by an enclosed overhead walkway, and of course the infamous refreshments rooms. When the GWR had reached the town, the company's directors entered an agreement with a builder in order to save some money. Messrs Rigby of London agreed to build the station and Railway Village at their own expense and in return were to be reimbursed from the tenants' rents from the houses and the takings from a refreshments room. They stipulated that every train must stop at Swindon for ten minutes, which suited the needs of Gooch and Brunel to change locomotives, but it created a monopoly on catering for the entire length of the railway. The result was a gold mine for the caterers and a recipe, so to speak, for culinary disaster for the long-suffering travellers. On one occasion Brunel was provoked into writing one of his vitriolic letters:

> I assure you Mr Player was wrong in supposing that I thought you bought inferior coffee. I thought I said to him I was surprised you should buy such bad roasted corn. I did not believe you had such a thing as coffee in the place; I am certain I never tasted any.

This compulsory halt was finally abolished in 1895. The old station was damaged by fire three years later, but it continued to serve until the 1970s when it was bulldozed to make way for an office block and new station buildings.

Thankfully, the Railway Village with its 'company houses' has survived. There is some confusion about who designed the houses; was it Brunel or his collaborator Matthew Digby Wyatt? I asked Felicity Jones about this: 'The Railway Village was very much Brunel's conception. He obviously had his own ideas on what he wanted, but it was Digby Wyatt who did the work.' Either way, this is a marvellous example of early planned industrial housing. Built between 1841 and

Brunel laid out the Railway Village, consisting of 300 cottages, in collaboration with Matthew Digby Wyatt. Constructed between 1841 and 1865, some were built with the excavated stone from the Box Tunnel.

1865, the 300 houses are laid out on a symmetrical pattern in neat rows along six parallel streets, all named after GWR stations. These 'company houses' were for the most part constructed from the mellow Cotswold stone hewn from the Box Tunnel excavations. Some look across the street and straight on to the Works' buildings, although none is more than a few hundred yards away. Closer inspection reveals that they are not all of the exact same size and design and the larger ones were occupied by the under-managers, foremen and their families. Even now you can appreciate that their former company inhabitants enjoyed a remarkably pleasant environment and standard of housing. Small front gardens create space between the rows and, behind the houses, Victorian Gothic arches open into long alleyways leading to the walled backyards.

The bulk of the Railway Village properties were purchased by Swindon Borough Council in 1966 in order to protect them. Lovingly preserved, they have changed little over the years, except perhaps for the removal of the iron railings along the front gardens – presumably during the war – and the old gas lamps. One of the houses, 34 Faringdon Road, is maintained and furnished by the Steam museum as an example of a 1900s worker's home, and this is open to visitors by appointment. It contains a small kitchen with its original cooking range, tin bath and mangle, a front parlour, and bedrooms with brass bedsteads and marble-topped washstands.

Working for the GWR has been described as more of a way of life than a job. Generations followed their fathers into the works, benefiting from a sense of security and, for most of the company's history, a job for life. There was also a great sense of community and loyalty and the company did much to take care of its workforce. At the centre of the village is the Mechanics' Institute building of 1855, which served generations of railway workers as the hub of their social activities. In addition to an auditorium, there was a reading room with magazines for the ladies and more serious material for gentlemen. There were also schools, shops, St Mark's church and a 10-acre park where workers and their families watched cricket

and rugby matches, attended fetes or listened to the various company brass bands. Many looked forward to the annual holiday week when almost the whole town – up to 25,000 people – boarded special trains heading off to seaside towns such as Weymouth or Weston-super-Mare. Their physical needs were taken care of at the GWR Medical Fund building, which included Turkish baths and two swimming pools. And in return for a small weekly payment, workers and their families had access to doctors' surgeries, dental and eye clinics, a casualty department and all manner of health care.

Situated on the edge of the Railway Village, in Faringdon Street, is a prominent twin-spired Gothic building that looks out across the busy road to the shops. This was built in 1849 as a lodging house to ease the accommodation shortage for single young men, but with its strict regime and cell-like rooms, it became known as 'The Barracks'. It served as a Wesleyan Methodist chapel for many years and in the 1960s was converted into a somewhat cramped home for the GWR Museum until the opening of the new one in 2000. Across the road and a short walk to the Brunel Shopping Centre – with its familiar collection of high street names – you reach an excellent statue of the man himself, standing tall on a column shaped much like the hat, which is otherwise conspicuous by its absence. Overlooked by the majority of passers-by more intent on getting their dose of retail therapy, a few still appreciate its significance and remember that this was once a railway town.

Swindon is also the start of the branch of Brunel's Cheltenham & Great Western Union Railway, which went via Cirencester/Kemble, the Sapperton Tunnel and through Stroud to Gloucester and Cheltenham. For more on this line, see *Brunel in Gloucestershire* by John Christopher.

In the middle of the Brunel Shopping Centre, the figure of IKB rises above the Saturday shoppers as a reminder of Swindon's roots as a railway town.

STEAM – MUSEUM OF THE GREAT WESTERN RAILWAY

LOCATION: Kemble Drive, Swindon, SN2 2EY.

GETTING THERE: By train to Swindon station, then a short walk. If driving, look for signs to 'Outlet Centre' or 'Great Western Heritage Area' or the 'M' for museum. The big car parks for the outlet centre are expensive.

OPENING TIMES: Daily 10.00–17.00 (closed 24–26 December and 1 January).

This is a great museum on the site of the former railway works. In the 'Building the Railway' section you come face to face with IKB, hear from the navvies who did the hard work, and admire the imposing *North Star* replica. There are pieces of the South Devon atmospheric pipe. When you have finished exploring the museum, take a look around outside and try to imagine what the 'Works' must have been like in its heyday.

INFORMATION: 01793 466646 / www.steam-museum.org.uk

SWINDON RAILWAY VILLAGE

LOCATION: Swindon, Wiltshire.

GETTING THERE: On foot, through the tunnel under the railway to the village area. Parking is virtually impossible in this part of town.

Working with the architect Matthew Digby Wyatt, Brunel created a village of 300 cottages arranged in neat rows, built with stone from the Box Tunnel excavations. Note the central alleyways between the cottages, the three pubs, Mechanics' Institute and former Methodist chapel which, until 2000, was the overcrowded home of the GWR Museum.

BRUNEL STATUE

LOCATION: Brunel Shopping Centre, Swindon.

GETTING THERE: Walk from the Swindon Railway Village.

It is a short walk into the Brunel Shopping Centre, where the statue looks down from a high industrial-style column. It is a copy of the Baron Marochetti statue in London, shown on page 73.

6

SWINDON TO BRISTOL

Westwards to Chippenham

In terms of construction, the GWR was very much a railway in two halves, with the easier gradients between Paddington and Swindon giving way to the more undulating terrain further west. The latter involved the greatest concentration of heavy work on the entire line, with extensive embankments, viaducts, assorted bridges and tunnels, including the longest railway tunnel ever attempted. At the outset, the thirteen-mile section between Wooton Bassett and Chippenham includes four deep cuttings and three high embankments, one rising to 40 feet. Construction of these was hard going and the poor winter weather of 1839 caused frequent slippages of the heavy clay. In some places, Brunel had wooden piles driven down either side of the embankments, lashed together with chains carried through the banks. The first train to travel from Paddington to Chippenham – a run just short of the 100-mile mark – pulled into the station on 31 May 1841. A local newspaper reported:

> The day was exceedingly fine and we are informed that the various trains throughout the day were much crowded, and formed great objects of curiosity to the inhabitants of the vicinage.

In some cases, curiosity gave way to outright fear as another eyewitness, the Revd Charles Young, recorded:

> … the strangeness of the sounds, the marvellous velocity with which engine, tender and trucks disappeared, the dense columns of sulphurous smoke, were all too much for my simple dominie [his parish clerk], and he fell prostrate on the bank-side, as if he had been smitten by a thunderbolt!

Located within an open site, Chippenham station's single-storey building of mellow stone is in Brunel's much-favoured Italianate style. With the opening of branch lines to Salisbury and Weymouth in the mid-1850s, the station was enlarged and an island platform and train shed were added (demolished in 1905). Further platforms have been added and the main down line is now on the south side of the island. Chippenham's agricultural trade and local industrial activities made it an important destination for goods traffic and a 194-foot-long goods shed was erected, which later made way for additional parking for the growing hordes of commuters. Entering from Station Hill, look out for the detached building on the left-hand side; a plaque states that this was the site office used by IKB during construction of this part of the railway. Two footbridges take you to the central platforms or out to the large north car park.

Bourne depicted the Chippenham viaduct – complete with herds of sheep and cattle – as if it were the gated wall of some ancient city. Today it still stands tall and proud, albeit as an over-sized traffic island.

Standing beside the front of Chippenham's station, this modest building is said to have been Brunel's site office during construction of this section of line.

Part of the appeal of the station today is that it has avoided the advances of modernisation, but this isn't Chippenham's most striking railway landmark. It is overshadowed by the impressively solid-looking viaduct which thrusts through the town like the fortified wall of an ancient city. Bourne depicted it as a romantic gateway complete with cattle, sheep and wagons passing through, and he described the viaduct as being 'extremely plain, but its effect, as seen from the road, is particularly good'. It is constructed with severe black bricks on the side facing into the town where some of the houses almost shelter under its arches, while creamy Bath stone faces outwards. Paradoxically it is so big, so 'in your face', that the locals hardly notice it anymore except as a huge traffic island, and if Bourne tried to sketch it nowadays he would probably get run over.

Box Tunnel

Work on the railway west of Chippenham leading to Box Hill commenced in 1837. With little of the line at ground level, there were two miles of high embankment followed by three miles of deep cutting through Corsham and into the eastern end of Box Tunnel. Box was a truly massive obstacle and typically IKB took the decision to drive straight through it with two miles of tunnel excavated in strata of hard oolite limestone, clays and soft fuller's earth. It also had the steepest gradient on the line, sloping downwards from Corsham towards Bath at 1:100.

Undaunted by this immense engineering challenge, Brunel had to contend with claims that Box Tunnel was a step too far. Even George Stephenson had expressed his doubts when the GWR Bill had gone before Parliament: 'The noise of two trains passing each other in this tunnel would shake the nerves of the assembly. I do not know such a noise. No passenger would be induced to go twice.' Ironically the elder Stephenson had been called as a witness in support of the GWR, and had only made these comments under cross-examination. One vociferous critic who needed no excuse to vent his feelings was the irrepressible Dr Lardner, who calculated that if the brakes were to fail, then a train would accelerate in the tunnel to over 120mph and the passengers would be suffocated! IKB pointed out that the good doctor had failed to allow for the slowing effects of air resistance or friction.

Construction on the tunnel began with the digging of eight vertical shafts into the Wiltshire hillside – the outer two later became part of the approach cuttings and one other was blocked off. Approximately 25 feet in diameter, they varied in depth from 70 feet to almost 300 feet down to the level of the line (these shafts still serve to ventilate the tunnel but they are all situated on private land with no public access). From the bottom of each shaft, the navvies began excavating in both directions in the hope that the sections would eventually join up. The only way in and out was via the shafts; excavated material was hauled up by horse gins, and the workmen and materials were lowered down. As the shafts were the only source of fresh air, conditions in the tunnel must have been hellish. Faced with hard limestone, the miners hammered holes into its surface with chisels, packed them with gunpowder and then lit the fuse to blast a way through. Within the confined space every blast and hammer blow was deafening and the result of the explosions was a shower of stone, clouds of dust and foul sulphurous smoke. Where the stone gave way to clay or fuller's earth, underground streams flooded into the works in many places. To cap it all, this dangerous and difficult work was undertaken with only the feeble light of flickering candles. Each week, a ton of locally made candles and a ton of gunpowder were consumed.

It was only when Box Tunnel was completed in June 1841 that trains could finally run all the way from London to Bristol. The two-mile tunnel is so straight that you can see the dot of light from the far end, but the enduring myth that the early morning sun only shines straight through it on IKB's birthday owes more to coincidence than to his design or to any other intervention. Even when completed, fears that the tunnel was dangerous to travellers were not easily dispelled and a

Bourne's interior of the Box Tunnel. Note the signalman standing to the right of the track, and also the dot of daylight visible at the far end.

local coaching company kept in business by taking nervous passengers over the hill to rejoin the train on the other side.

Box Tunnel had been a colossal feat of engineering and hard labour, which had taken an appalling toll, with upwards of 100 workmen losing their lives. IKB has sometimes been criticised for his almost callous attitude to the human cost in creating his great works, and when asked by a parliamentary committee about the fatalities, he expressed surprise

Sydney Gardens in Bath is an unlikely setting for a railway cutting, but the strong curves of the wall combined with Brunel's elegant bridges and stonework make this one of the most photographed locations on the whole GWR.

that the number should be so low. But in his defence, it should be remembered that his treatment of the workmen was no worse or better than any other employer. To his credit he had enormous admiration for what they achieved and they, by all accounts, had great respect for him. When the tunnel was found to be perfectly aligned, it is said that he took a ring from his finger and presented it to one of the foremen.

The most famous aspect of Box Tunnel is undoubtedly its elegant western portal, which can be seen from the busy A4 or the new viewing area. Unfortunately, the other entrance is situated at the end of a long cutting and is neither easily seen nor very attractive, as it has been extensively modified

over the years. In addition to digging the tunnel and cuttings, excavations continued at Box for building stone and by 1864 around 100,000 tons were being transported from Corsham station annually. On the north side of the eastern portal, there is the entrance to the tunnel quarry, which burrows deep into the hillside. During the Second World War, this labyrinth of underground chambers was utilised as an ammunition store handling up to 2,000 tons of ammunition daily, although it was not permitted to store explosives within 100 feet of the railway tunnel.

Just a little to the west of Box is the Middle Hill Tunnel, which has a fine stone portico, although it is difficult to access for a close-up view.

Bath Spa

Having survived the perils of Box Tunnel, the railway traveller soon enters the elegant town of Bath, where the line passes under the Beckford Road bridge and then runs smack through the middle of Sydney Gardens. This is one of the finest and most photographed railway cuttings in Britain and Bourne described it as 'the Vauxhall of the place'. Far from spoiling the scene, the locals themselves pronounced that the railway had 'been arranged so as to increase, rather than injure, the attractions of the gardens'. Aesthetics aside, the main engineering problem entering Bath was to divert the Kennet and Avon Canal on the slope above the line. For the railway, IKB created a cutting with a strong retaining wall of stone on one side and faced on the other by a remarkably low balustraded wall. Until only recently there was no security fencing on the park side and as the London-bound trains hurtled through on the nearer line, they seemed so close that the thunderous noise and sudden rush of air left you shuddering. Brunel designed two elegant pedestrian bridges to cross the cutting; one in stone and the other made of iron. At the end of the nineteenth century, there was a temporary station in Sydney Gardens itself while the main Bath Spa station was being improved. Unfortunately, the impending electrification of the London to Bristol line will entail the installations of gantries within the cutting. A great shame, as this will significantly clutter up the view of this photogenic cutting.

Heading deeper into Bath, the trains disappear under Sydney Road and are engulfed by two short tunnels running beneath the houses. At the base of the first archway, there is a plaque erected by the Brunel Society to 'Isambard Kingdom Brunel Engineer 1841' – a superfluous adornment in such a distinctive location. Beyond the tunnels, the elevated line crosses Pulteney Road (A36), carried on a bland 1970s steel span in place of the original bridge, and continues along the imposing thirty-one arches of the curving Dolemeads viaduct leading to Bath station, which nestles within the elbow of the River Avon. The approach from the east is over the 88-foot span of the St James's Bridge, a fine stone structure so peppered with a patchwork of brick repairs that it is hard to be sure what is left of the original.

From the outside, Bath Spa station is an oddity. It sits amid this famously formal Georgian town looking for all the world like a Tudor farmhouse. Step inside and you discover that this two-storey building is something of an illusion, with the elevated track on the upper level. Having climbed up to the platforms, you can make out the distinctive Brunellian style of door and window detailing. Originally Bath Spa had a full all-over wooden canopy covering both tracks, and Bourne depicted an elegant structure on a par with Brunel's Temple Meads in Bristol. That was replaced with the later canopies and these obscure the bigger view of the station building.

Behind the station, a later footbridge takes you over to the riverside towpath. Head left and you reach the entrance lock to the Kennet and Avon Canal and along to the St James's Bridge. Turn right and you pass under the steel girder railway bridge that in 1878 replaced IKB's wooden Skew Bridge

Left and opposite: This steel girder bridge replaces Brunel's wooden bridge, which crossed the Avon at a sharp skew on the west side of the station.

crossing the Avon at a very oblique angle. Beyond this, the striking crenulated viaduct over the A36 Lower Bristol Road still survives and marks the start of the long Twerton Viaduct heading west out of Bath.

Continuing westwards, the line passes through a series of tunnels – some in a romantic style – at Middle Hill, Twerton and Saltford, plus the three known as No.1, No.2 and No.3, of which the latter is over 3,000 feet long. There are the Twerton and Deep Ashley cuttings, and several bridges. The most notable of these are single-arch bridges over the River Avon at Bathford and, as the line approaches Temple Meads, it crosses over the Avon again, via a fine masonry bridge. J.C. Bourne depicts this bridge with one main span and two smaller side spans, striding over the river amid open countryside, but today the area is entirely built-up and the bridge remains, although almost entirely hidden by later lattice girder bridges added on either side. Finally, the line reaches the meadows or 'meads' at Temple Meads. It was here that Brunel's unfettered imagination conjured up a grandiose terminus worthy of his Great Western Railway.

The fantastic facade to the world's oldest surviving railway terminal, Brunel's glorious Temple Meads. Carriages would enter through the arch on the left – note the circle where a clock once kept standardised railway time – pass under and through the building to exit via a matching arch on the right (now gone). The company directors met in offices upstairs.

There is great attention to detail on the buttressed and crenulated viaduct just to the west of Bath station. Note the coat of arms and the loco wheel motif, but try to avoid the busy traffic!

Bristol: Temple Meads

Brunel's 'old station' is situated to the left of the Station Approach Road, with its frontage of Gothic-Tudor stonework facing out on to the Temple Gate A4 road. The GWR's coat of arms can still be made out at the top of the building, and it was here that the offices and directors' boardroom were located. Passengers arriving by carriage or on foot entered the station through a pair of arches on the left-hand side. There was a clock over these because the coming of the railways had created a need for standardised time across the country if passengers were to catch their trains. Carriages passed through the lower level of the station to the booking office and waiting rooms, where steps led up to the platforms (they exited through a matching pair of arches on the right of the building, now demolished). Inside the passenger station or shed, travellers were astounded by the loftiness and lightness of the building. Bourne described it as being:

... occupied by five lines of railway, covered by a timber roof of 74-foot span, pierced with skylights and ventilating funnels, and in a style of architecture suited to the exterior of the building. It is separated by a row of iron columns and flat arches from the aisles, which are occupied as platforms for the arriving and departing passengers, and are lighted by a range of windows along the outer wall. The east end of the shed is open to receive the Railway ...

His engraving of the interior of Temple Meads is probably his most famous and it accurately portrays the distinctive mock-hammerhead beams; a purely decorative feature with no structural function. Beyond the colonnade of neo-Tudor arches running the length of the platforms were unglazed window openings for ventilation, and additional vents were later added at the apex of the roof for the smoke and steam to escape.

The engine shed was sandwiched between the passenger shed and the offices. As this was a terminus in the strict sense of the word, the engines were revolved on a turntable at the rear of the shed to point them back the other way. Beneath the tracks, there were openings leading to brick-vaulted cellars, and the ash and clinker from the engines would be raked out and swept through into trucks down below. Above the engine shed was a large water tank, which can be seen rising like a tower near the front of the building.

In order to handle increasing rail traffic, the new Temple Meads building was constructed between 1871 and 1878 to a Gothic style widely attributed to IKB's old collaborator Sir Matthew Digby Wyatt, and the old station remained in use as Platforms 13 and 14 until 1965.

I first visited Brunel's Temple Meads back in the early 1980s when I undertook a photographic survey of his work in Bristol. It was in a very sorry and neglected state, cut off from the rails by later buildings and relegated to a humble car park. By 1983, the 150th anniversary of IKB's appointment as engineer to the GWR, the Brunel Engineering Centre Trust had stepped in to rescue it. Paying a peppercorn rent to British Rail, they were responsible for the upkeep of the fabric of the building – a task comparable in scale to the restoration of a cathedral. There was much discussion concerning what to do with the space and initially it was proposed to make it into a national centre for civil engineering. That scheme fell by the wayside and the passenger shed now serves as a venue for events but sad to say it is not normally accessible to the public.

Bourne's engraving of the inside of Temple Meads shows five lines of broad gauge track. The windows to either side were unglazed and further ventilation was later added to the apex of the roof for steam and smoke to escape.

Even so, two of the roof bays complete with unrestored hammerhead beams can be viewed from the entrance to the later red-brick extension currently used as covered parking.

For a while, the engine shed became much more accessible as the home to the Empire & Commonwealth Museum, which opened in the 1990s, but this has since closed.

Temple Meads is an incredible survivor. The oldest purpose-built railway terminus in the world – pre-dating Paddington by over twenty years – it has been designated as a Grade I-listed building, which is a fitting accolade for one of Bristol's most important landmarks.

Cut off from the rails, by the 1980s Brunel's Temple Meads building was in a sorry state and served as a covered car park.

Passengers marvelled at the 74-feet span of the wooden roof, the widest in the world at the time. The hammerhead beams, shown here after restoration, may look impressive but they are purely for decorative effect.

CHIPPENHAM VIADUCT AND STATION

LOCATION: Chippenham, Wiltshire. (OS map 173:987736 and 921737)

GETTING THERE: Main line train. From the M4, take junction 17 south on A350. From east or west on the A4, park in the town or at the two large car parks either side of the station.

The viaduct at Chippenham strides across the town – the outer side looking like the fortified gateway to a walled city, while the inside is all dark brickwork. Enter through the archway into New Road, where a left turn takes you up the slope of Station Hill. Approaching the booking hall, note the small detached building on the left, said to have been Brunel's office during construction of the railway. You will also see two footbridges – one within the station that connects with the central platforms and the far side of the track, while the other leaps over the whole railway without entering the station.

BOX TUNNEL

LOCATION: The famous western portal is situated at the edge of Box Village, five miles east of Bath on the A4 to Chippenham. (OS map 173:829688)

GETTING THERE: The A4 is very busy and there is no public parking, so best to park in Box and walk up to the tunnel.

There is a small viewing area complete with plaque, although the road bridge provides a better view. There is no pavement here, so you need to take care. Proceed another 100 yards up the hill for access via stiles into Lacy Wood, a triangle of land beside the cutting. This is the Box Millennium Project and the closest you can get to tunnel, but the view is restricted by summer foliage. By contrast the eastern portal, emerging into a long cutting two miles away on the other side of the hill, is hard to find and very disappointing. You can walk along the footpath beside the cutting, but a new security fence keeps you too far back to see anything. Note the overgrown pillboxes, put there to defend the tunnel and ammunition depot in the Second World War. (OS map 173:858694)

At the crest of the hill between Box and Chippenham, take a moment to look down the valley and to the city of Bath in the distance. Stunning.

BATH SPA STATION AND SYDNEY GARDENS

LOCATION: Central Bath, BA1.

GETTING THERE: London to Bristol main line, or take A36 into Bath. Parking at the station is limited and very expensive. There is a multi-storey car park on the other side of the bus station. It is far better to take advantage of the Park and Ride scheme from locations around the city's perimeter.

The railway passes directly through Sydney Gardens – a lozenge-shaped park surrounded by the A36 less than a mile from the station. It is one of the most picturesque railway cuttings you could wish to see, with the Kennet and Avon Canal a little higher up the slope. On-street parking is possible on Sydney Road, but can be limited.

Bath station sits within the curve of the river, and Brunel chose an Elizabethan style for this prominent building within this famously Georgian city. Because the railway line is raised, the station building seems low when viewed from the platforms and it is hard to see behind the canopies (it originally had an all-over roof). Note the wonderful GWR benches. Behind the station, the river is crossed via a footbridge. Alternatively come out of the main station building and head left along Dorchester Street to the A36 Lower Bristol Road, where the crenulated arches of the long viaduct form a roundabout. The bridge west of the station replaced IKB's wooden structure. Follow footpath beside the river to reach Bath Locks and the much-patched St James's Bridge.

TEMPLE MEADS RAILWAY STATION

LOCATION: Station Approach, Temple Meads, off A4 Temple Gate, BS1 6QF.

GETTING THERE: By train is easy. Parking near this busy station is not.

There is no access for trains to Brunel's old station building. The former GWR offices and boardroom are situated in the upper floors of the Elizabethan-style building. The left-hand archway served as the entrance for passengers in their horse-drawn carriages, but a matching exit arch on the other side has gone. The restored passenger terminus now serves as a conference and event venue. Note the twin-towered stone building on the other side of the station approach, the original Bristol & Exeter Railway's building.

7

CLIFTON SUSPENSION BRIDGE

Bristol: Chain Reactions

Location, location, location. Isambard Kingdom Brunel may not have invented the suspension bridge, but to span the spectacular Avon Gorge at Bristol, he came up with one of the finest the world has ever seen. As L.T.C. Rolt put it:

> Today, in a world satiated with engineering marvels, the grandeur of Clifton Suspension Bridge can still uplift the heart. More surely than any lineaments in stone or pigment, the aspiring sight of its single, splendid span has immortalised the spirit of the man who conceived it …

It was this bridge that launched IKB's meteoric career and cemented his relationship with Bristol, the city most closely associated with his work. Even so, it is not entirely certain what brought him here in the first place. It might have been part of his recuperation following the Thames Tunnel flooding, but there appears to be no documentary evidence to support this. More likely he was drawn to the city because of the competition to design a bridge and it was the subsequent chain of connections that resulted in such a concentration of his work in one place. Indeed Bristol often claims IKB as an 'adopted son', but the feeling may not have been mutual. Significantly, IKB never had a permanent home in the city and preferred to live in London. What

Bristol did do was to provide him with opportunities by the bucketful, and none could have been timelier than the bridge competition.

The first proposal for a crossing over the Avon Gorge appeared more than fifty years before Brunel was born. In 1754 a prosperous wine merchant named William Vick hit upon the notion, which was both remarkably visionary and curious, as there was very little on either side of the gorge to merit such an ambitious link. On the Somerset or Leigh Woods side there was just that, woods with only the odd house, while on the other side the fashionable suburb of Clifton had yet to be built. Despite this, Vick got the ball rolling by bequeathing the sum of £1,000 to be invested until it grew to £10,000, when it was to be used for the construction of a stone bridge.

By the turn of the century, the elegant terraces of Clifton had indeed become the place to live for the merchants who were enjoying Bristol's new-found prosperity. Suddenly Vick's bridge was back on the agenda and in 1793 the appropriately named William Bridges published a grand design for an elaborate stone-built bridge. This multi-storey structure consisted of five layers with houses, a granary, corn exchange, chapel and tavern, surmounted by a lighthouse and pierced at its base with a high archway through which vessels would pass. More of a vertical village than a bridge, it was hugely ambitious and impossibly expensive. Given our familiarity with the bridge that spans the gorge today, it is all too easy to dismiss this Munchausenesque fantasy out of hand. But – and I'm going out on a limb here – I would suggest that if

The Avon Gorge, a fantastic setting for Brunel's 'first darling' – the Clifton Suspension Bridge. The entrance to the docks can be seen at the bottom right-hand corner of the picture. (Derek Maltby)

Originally sited at the end of Narrow Quay, John Doubleday's statue has been relocated to Temple Quay near Temple Meads.

this preposterous design had been built, then we would have come to revere it just as much.

Interest in the bridge scheme languished in the doldrums when Bristol experienced a sudden downturn in its commercial fortunes, due in no small part to the economic uncertainty caused by the war with France, and there was little thought for such a venture for the next twenty years. It was only once Napoleon had been dealt with that civic confidence returned. By then the engineering marvels of the Industrial Revolution were beginning to make their mark on the nation's landscape and the city fathers wanted a slice of the action. Vick's legacy had grown to around £8,000 and a committee was formed to get the job done, but they soon realised that it might cost ten times that much. And while Vick had stipulated a bridge of stone, the rapid advances in new materials, especially wrought iron, meant that an iron suspension

bridge was a far more affordable alternative. Accordingly a competition to design a bridge for the Avon Gorge was announced on 1 October 1829, with the lure of untold fame and glory plus a cash prize of 100 guineas. The response was phenomenal and twenty-two sets of plans were submitted by late November, with the young Brunel among the aspirants.

IKB had endured a period of dark despair following the cessation of work on the Thames Tunnel. This hiatus was made all the more gruelling as he watched his contemporaries – especially his close friend Robert Stephenson – making names for themselves while he was occupied with only a handful of minor commissions. The bridge competition gave him the opportunity to kick-start his career and characteristically he set about the task with enormous energy and commitment, plus a little help from his father. He studied suspension bridges throughout the country, including Thomas Telford's 1826 road bridge over the Menai Strait, before submitting no less than four designs. Each had a main span of around 900 feet, which was far in excess of any seen before. Two featured tall towers, while the third showed both approaches buried in tunnels through the limestone cliffs and chains anchored into the rocks above. His fourth design, known as the 'Giant's Hole' design, and reputedly the one he most favoured, had a tunnel on the Clifton side which emerged in the proximity of the existing cave of that name.

To judge the entries, the Bridge Committee called upon the services of Thomas Telford, but unfortunately his confidence in suspension bridges had been severely shaken when his newly completed Menai Bridge had encountered problems with wobble. The effect of side winds upon a bridge was little understood then and, although the wobble had been cured, Telford had become much more conservative in his approach. He rejected every single submission to the Clifton

competition and decreed that 600 feet was the absolute maximum span for a suspension bridge. And who was going to disagree with the great man? Well, IKB for one! Matters went from bad to worse when Telford was asked by the Bridge Committee to produce his own design and this – surprise, surprise – was much admired and forwarded with the Bill for parliamentary approval in 1830.

In order to keep the span within his self-imposed limitations, Telford's design featured two ornate Gothic stone towers, rising upwards 200 feet from the floor of the gorge like needles, supporting a central span of 600 feet with shorter approach spans to either side. The Bridge Committee liked it, but the public did not. When the flow of money from investors proved inadequate for its construction, the committee discreetly put aside Telford's design and a second competition was announced, this time for a bridge costing £40,000. No doubt Telford was a bit miffed, but spare a thought for W. Hawks. His design actually won the second

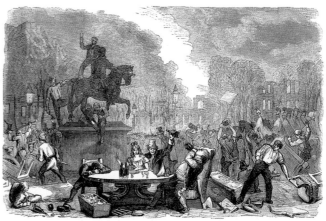

Queen Square is now an oasis of calm and quiet, but during the Bristol Riots of 1831, many of its fine buildings were razed to the ground. Brunel was enrolled as a special constable during the disturbance. William III can be seen on the horse.

competition, prompting IKB to descend upon the judges with a tactic that he would deploy many times throughout his career. They had placed his entry in second place, having expressed concerns over the chain design, anchorage and suspension rods. His response was to bombard them with science, reams of drawings and calculations, all the while cajoling them with a charm offensive. As a result he actually succeeded in persuading them to reverse their decision and to accept his design, which became known as the 'Egyptian thing'. This featured a single span slung between sturdy supporting towers, or piers, with the one on the Leigh Woods side perched on an abutment to shorten the central width to 630 feet. The towers were decorated in the Egyptian style popular at the time, each capped with a pair of sphinxes – initially facing into the gorge, but later turned around so that approaching carriages would not be greeted by the sight of their bottoms. Brunel was cock-a-hoop at his own cleverness. If only the construction of the bridge, his 'first child' as he affectionately called it, could have gone to plan. But he was not to witness its completion in his lifetime.

Almost from the word go there were problems. On 21 June 1831 a ceremonial start was made amid much pomp and speech making, but within a matter of days Bristol was in chaos, torn by riots in opposition to the proposed Reform Bill. Never one to miss out on a little excitement, IKB enrolled as a special constable, and as the mob ransacked the Mansion House in Queen Square, he brought his full 5 feet 3 inches to bear in protecting the city's treasures – even arresting one of the rioters. As a result of the ensuing unrest and financial uncertainty, further subscriptions to the bridge scheme dried up and it was another five years before work commenced. On 27 August 1836, the first stone was laid for the Leigh Wood's abutment – of red sandstone quarried locally – and by 1840 the abutment was completed. At a cost of over £13,000, this was an expensive monument to Telford's folly as it had only been introduced by IKB to bring the span nearer to the old man's 600 feet. By 1843, when both towers were ready to receive the chains, it was announced that the fund of £45,000 had been exhausted. Unfortunately it soon became apparent that sufficient further funds could not be raised, partly because of the arrival of the GWR, which was a more attractive investment to the Bristolians. With work halted, it was decided to sell off the chains and despite IKB's best efforts at procrastination, they were sold in 1851 to be used on the Royal Albert Bridge at Saltash. This was the death knell for his beloved Clifton Bridge, and the towers standing tall and proud on the cliff tops

The opening ceremony, as depicted by the *Illustrated London News*.

remained as a constant reminder of unfinished business. Some locals even campaigned for their demolition.

Still, at least they provided some amusement for a few thrill-seeking Bristolians. When work on the bridge had started in 1836 an iron bar, 1,000 feet long and 2 inches in diameter, had been slung across the gorge and fixed into masonry on either side. If you look over the parapet on the Leigh Woods abutment you can still see the stump or plinth for the cable among the undergrowth. During construction of the bridge, the bar provided a convenient means of transporting workmen and supplies from one side to the other via a basket suspended underneath. Gravity pulled the basket to the middle and it would be hauled up the other side by a rope. This bar had been left in place and local daredevils would pay the not inconsiderable sum of five shillings for the dubious pleasure of travelling across the gorge by this means. Inevitably there were misadventures and, on one occasion, a newly married couple found themselves stranded halfway after the rope had parted from the basket, and there they stayed until a new rope could be passed to them. They were in good company, for it is said that IKB had once got stuck in the same place.

Completing the Bridge

Ironically it was Brunel's death in 1859 that provided the impetus to get the job finished. The Institution of Civil Engineers decided that the Clifton Bridge should be completed as a monument to their former colleague, and by chance the chains from the Hungerford Bridge had become available at just the right moment. In May 1860 a new bridge company was formed and the following year, Parliament passed a new Act for its completion. Work resumed in 1862 and two years later, on 8 December 1864, the opening ceremony took place.

The bridge we see today varies in several ways from Brunel's design. Most obvious is the lack of the sphinxes and also the Egyptian-style cast-iron panels, which were to adorn the towers with scenes of the bridge's construction. On the span itself, a third chain was added to form a top layer above the Hungerford double chains. The chains are made up of sections with alternately ten or twelve links of flat metal placed side by side and bolted together. For the deck, IKB had specified a timber framework with iron straps, but instead it was constructed with wrought-iron plate girders and lattice cross girders. The main strength of the deck comes from two girders running the full length of the bridge and these are visible as dividers between the roadway and the footpath. The deck's width was increased from IKB's 24 feet up to 30 feet – which is just as well, given modern traffic. To construct it, 16-feet sections were lowered into place by cranes and the suspension rods were then attached, 162 in all, varying in length from 3 feet to 65 feet.

The completed bridge was load-tested with 500 tons of stone ballast and it was found that the middle of the roadway sagged 7 inches, well within acceptable tolerances. In fact the bridge moves up and down all the time, and also sideways in strong winds. In higher temperatures it expands and the chains are approximately 3 inches longer on a hot summer's day compared with the cold of winter. The road level is slightly higher on the Clifton side, by 3 feet in fact, and IKB deliberately incorporated this into the design to create the appearance of the bridge being level. At either side, the chains run over rollers at the top of the towers and down to ground level, where they are deflected by saddles into excavations 60 feet deep into the rock. They were originally anchored underground by a system of Staffordshire brick in-fills, which spread them out like a wedge, but since the 1920s they have been held in place by 20 feet of solid concrete. Which is reassuring as the bridge weighs in the region of 1,500 tons and carries loads of up to 4 tons at peak times. Without this anchorage, the towers would be pulled in towards each other.

Somebody who knows this bridge intimately is the former Bridge Master John Mitchell, who has been involved with it since the 1970s, becoming Bridge Master in 1996. I asked him if the bridge had a character and whether it commanded affection in the way that a ship might from its crew. 'Crikey – yes, I have great affection for the bridge,' he responded thoughtfully. 'She does have a character. She can be so beautiful and pleasant, but she has a dark side too.'

I know what he means. As you step on to the bridge, you can't fail to notice prominent signs for the Samaritans. Understandably it's not something the bridge people want to talk about, but since its opening this high structure has attracted many with suicide in mind. Nowadays the side railings are much higher than they used to be and the bridge staff are always on the lookout for any telltale signs among the stream of pedestrians. Over the years there have also been some remarkably close escapes. In 1885, a young woman named Sarah Ann Henley jumped from the bridge and her wide skirts billowed out, acting like a parachute and carrying her relatively gently down to the water. In 1896, two girls named Elsie and Ruby Brown were hurled from the bridge by their father, who was said to be suffering from 'temporary derangement'. Maybe it was their young age combined with a high tide, but they survived the fall and were rescued by the pilot boat. There was also the strange case of Lawrence Donovan, who was not suicidal but claimed to be a high-diving expert and intended jumping from the bridge. An unimpressed toll collector told him to go away, but Donovan was persistent and one evening he returned with a band of helpers, who chucked a dummy off the bridge and yelled 'He's jumped!' Meanwhile their associates made a big commotion down below as they pulled Donovan out of the water. At the hospital he claimed that he had survived injury because he had protected himself with plates of metal; not to lessen the force of the impact, but because of the electricity they were supposed to produce. No one knows if the other dummy survived the fall.

The gateway created by the bridge and gorge has also acted like an irresistible magnet to some reckless aviators over the years, although this

A sketch of the secret chambers found within the Leigh Wood's abutment or pier in 2002. As no plans survive from its construction, the abutment was assumed to be solid.

practice is strictly illegal now. The first to pop under was Frenchman M. Tetard in 1911. He survived, but when Flying Officer J.G. Crossley of 501 Squadron took the plunge in 1957, he was killed when his Vampire jet smashed into the side of the gorge at 450mph.

Nowadays the Clifton Suspension Bridge is looked after by a trust established by a parliamentary Act in 1952. To cover its upkeep, drivers pay a modest toll while pedestrians and cyclists have crossed free of charge since 1991. This money pays the wages of the staff and covers the costs of repairs and maintenance. In less safety-conscious times, it was not uncommon to see the workmen clambering along the chains, but nowadays they are secured by harnesses or within a travelling cradle that moves beneath the roadway. Because of the bridge's exposed position, it needs constant protection from the elements. Originally its metal parts were coated with coal tar but they are now protected by a zinc coating beneath the paint.

It is remarkable to think that when Brunel designed the Clifton Suspension Bridge it was only intended for the use of horse-drawn carriages and he could not have anticipated the stream of motor vehicles that now cross it. It is estimated that around 3,000 commuters use it

daily and each year it carries 4 million cars. So I asked John Mitchell how long he thought the bridge might last – will it still be standing in 200 years, for example?

Well, if we put aside the worse scenario of say an earthquake, and we continue to manage the bridge as we do today, then it should last many hundreds of years. The most important components are the chains and the towers and this is why loading is important. We have a limit of 4 tons, or 600 pedestrians, and while we can manage the vehicles with barriers the pedestrians are more difficult.

That explains why the bridge is closed during big public events such as the Balloon Fiesta, which takes place at Ashton Court on the Leigh Woods side.

John Mitchell retired as bridge master in 2006, shortly after the Brunel 200 celebrations, for which he oversaw the installation of new lighting. Ever since its opening the bridge has been illuminated, at first with magnesium flares, which tended to blow out, then by

A fantastic photograph, taken during initial explorations of the cathedral-like spaces found within the Leigh Woods abutment. (Robert Lisney)

electric lighting, which was added in the 1930s, and since the Queen's Silver Jubilee in 1977 it has been floodlit. The new lighting consists of 3,000 tiny light-emitting diodes to illuminate the chains, fluorescent lighting for the walkways and additional low-level floodlights for the towers, abutments and underside of the roadway. Another development originally intended for completion in time for the celebrations was a long-awaited visitor's centre on Suspension Bridge Road. The trust used to have an exhibition in Bridge House, but this building was refurbished as apartments and in its place the trust has obtained planning permission for a purpose-built centre. The present visitor's centre is on the Leigh Woods side.

Clifton's Hidden Chambers

You would think that 150 years after the bridge was opened we would know everything there was to know about its construction, but in 2002 a chance find revealed its greatest secret. While drawings of the bridge have survived from its completion in 1864, there are none from the earlier period to show how the massive Leigh Woods abutment was constructed. I had often puzzled about this and because it would have been such a massive task to infill with solid material, I assumed that it must be hollow. The former bridge master, John Mitchell, was inclined to agree:

I thought it must have been built in the same way as a viaduct. Ever since it was built the face of the abutment has suffered from lime staining, indicating that water was carrying free lime. But we had no drawings. In the 1960s Sir Alfred Pugsley, who was a trustee, carried out several bore tests and found nothing. And when Sir Alfred says the abutment is solid, then it's solid! Incredibly the borings had missed their target, as did more recent probing by ground radar. Small internal chambers had been found on the Clifton side in 1978 and again in 1999. But it wasn't until a 3-foot shaft was uncovered during some paving repairs that the truth beneath the Leigh Woods abutment was finally revealed.

We assumed this shaft was to do with drainage and so we erected a tripod above it and one of our abseiling specialists, John Corber, went down to take some measurements. He was being lowered when I heard a sudden cry and a few choice words! I immediately looked into the hole, but he had gone and the rope was slack. Naturally I feared the worst. Then his feet suddenly reappeared and he shouted up that he had just put his head in a chamber 36 feet high!

John Mitchell was one of the first people to enter the cathedral-sized space and it was a moment he will always remember: 'It was absolutely astounding. What struck me was that there simply wasn't sufficient room to hold such a vast chamber! But this wasn't the only one. Through one hole and then another shaft and there was another chamber, even bigger than the first one.'

There turned out to be twelve chambers in all, with seven on the upper level and a further five below, all linked by a complex maze of narrow shafts. John's bigger chamber was directly under the road and had a length of 60 feet and height of 36 feet – about two-thirds of the height of the nave at Bristol Cathedral. Naturally each chamber is in darkness, although the air quality is good. Long stalactites stretch down spindly fingers almost to the ground. Known as 'candles', these are hollow strands of limestone barely a finger's width, and they could be broken by a breeze or even a careless shout. For John Mitchell, the discovery of this underground kingdom has left a lasting impression – initially of great elation and awe, only later balanced by the realisation of an enormous maintenance liability. Unfortunately the difficulties with safe access and the fragility of the stalactites mean that public access is not possible.

There are other curiosities to look out for on the bridge. In particular spot the differences between the two towers; the Clifton one has a much less pointed upper arch, its corners are square whereas the other has rounded-off corners, and it is the only one to have side apertures. In addition, nobody knows why there are several courses of lighter coloured stone on the Leigh Wood's tower or, for that matter, where to find the abutment's foundation stone.

In 2014 the Clifton Suspension Bridge celebrated the 150th anniversary of its opening in 1864. Of all the many inscriptions and plaques on the bridge, there is one up high on the Leigh Woods tower that stays with me. In Latin it reads, 'Suspensa Vix Via Fit' – a pun on the name of William Vick – and roughly translates as, 'A suspended way made with difficulty'.

Pedestrians crossing the Clifton Suspension Bridge.

The shaft on the Wapping side of the Thames Tunnel has been incorporated within the station building and its capped top can be seen above the booking hall.

The statue of Brunel at the Brunel Engine House, Rotherhithe.

Baron Marochetti's 1877 statue of Brunel overlooking the Thames Embankment.

The Charing Cross Bridge constructed on the piers of Brunel's Hungerford Bridge, as seen from the London Eye and looking towards Charing Cross station.

One of the two Golden Jubilee bridges that flank the railway bridge – note the brick piers, which have survived from Brunel's pedestrian bridge.

Delicate, almost art nouveau tracery at the 'country' end of Paddington station. Highly decorative, this screen also serves to stiffen the structure.

The directors' balcony above Paddington's war memorial, and entrance to the Royal Waiting Room.

The proposed Phase 2 redevelopment of Paddington station would have resulted in this modern glasshouse to accommodate London's new Crossrail. In the event, Span 4 was given a reprieve from the wrecking ball and Crossrail relocated to the western side of the station beneath Eastbourne Terrace. (Nicholas Grimshaw & Partners)

Sir John Fowler's 1920s facade of the Metropolitan railway station on the other side of Praed Street.

The former stationmaster's house at Steventon. Designed by Brunel, its central location was chosen as the weekly meeting point for the company's directors when the London and Bristol boards combined in 1842.

The Moulsford Bridge, which crosses the Thames at an angle of 45 degrees. Get up close to appreciate the incredible brickwork curves of the broad elliptical barrel arches.

The wide brick arches at Maidenhead – the critics had predicted that it would collapse under the weight of the trains.

A GWR publicity illustration of the *Cheltenham Flyer* crossing the Maidenhead bridge.

This imposing building in Faringdon Street, Swindon, began life as the notorious 'Barracks' – accommodation for single male employees – and in later life it served as a Methodist chapel before becoming the temporary and cramped home of the GWR museum until the 1990s.

The famous *Cornish Riviera Express*, on display at the Steam museum, epitomises GWR locomotive engineering at its best.

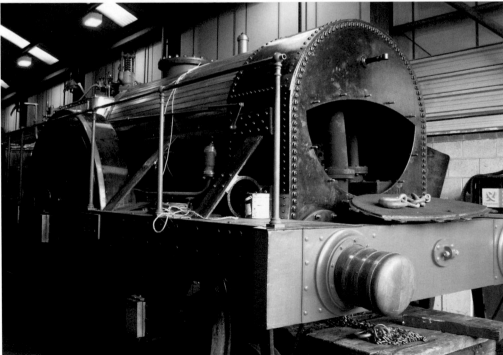

A working replica of Gooch's magnificent *Fire Fly* broad gauge loco at the Didcot Railway Centre.

Ironically, this full-sized replica of *North Star* at the Steam museum was built for the Stockton & Darlington's 100th anniversary of 1925 and was wheeled out again for the GWR's in 1935. Only the massive 7-foot diameter driving wheels remain from the original loco.

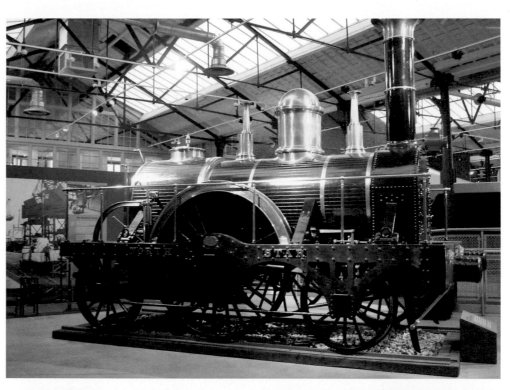

Brunel's samples of stone from the different areas of the GWR – his building blocks for the railway.

When the Box Tunnel was opened in 1841, it completed the link from London to Bristol. This is the more famous western portal.

The long viaduct on the eastern side of Bath Spa Station leads over the Avon, via the 88-foot span of St James's Bridge.

An Elizabethan farmhouse in the heart of this famously Georgian city, Bath Spa station is a conundrum with the track and platforms on its upper level.

When it opened, Bath Spa station featured a characteristically Brunellian all-over wooden roof; this was removed in the late nineteenth century and the newer canopies tend to obscure the original architecture.

A view of one the bridges that pass over the cutting at Sydney Gardens.

Bristol's most recognisable landmark, the
Clifton Suspension Bridge, was not completed
until after Brunel's death.

The tower on the Clifton side has these
slender apertures, unlike the one on the
Leigh Woods side. This is just one of
several differences between the two.

By the time the bridge was being completed, its iron chains had already been sold for the Royal Albert Bridge at Saltash, so replacement chains were acquired from Brunel's newly dismantled Hungerford bridge in London. A third top chain was added for extra strength.

The Great Western Hotel, Bristol, built to accommodate passengers travelling from London on their stopover before catching the boat to New York. The building was a Turkish baths at one time, but you wouldn't want to bathe there now, as it is currently offices for the planning department.

Mud! The result of Bristol's tidal waters and the bane of the Dock Company.

Brunel's South Entrance Lock into the Cumberland Basin is no longer in use. The tubular iron bridge is a replica, with the original in the background under the Plimsoll road bridge.

A contemporary print of 'The Iron Steam Ship *Great Britain*', under sail and in steam. (US Library of Congress)

Looking across Bristol's Floating Harbour to the Great Western Dock, where the restored *Great Britain* has become a new landmark and tourist attraction for the city.

Launched on 19 July 1943, the Great Britain returned to the same dry dock exactly 137 years later to the day.

A beautifully preserved example of a Pangbourne-type secondary station at Culham, on the Didcot to Oxford line. It is full of charming Brunellian features – note the signature skewed chimneys.

Brunel came to Stroud station in Gloucestershire in April 1845, when it opened on the Cheltenham & Great Western Union Railway, operating from Swindon via Cirencester to Gloucester.

Mixed gauge complexities at the Didcot Railway Centre. The outer rails are for broad gauge, while the inner rail reduces the width for standard gauge trains.

Up close and personal. The red cliffs and sea wall to the east of Dawlish make this stretch of line a firm favourite among railway enthusiasts and photographers.

Inside the transfer shed at Didcot, with broad gauge track on one side and narrow or standard gauge on the other. Passengers would make a quick dash between the two, but goods could be delayed for many hours in the changeover. (Ute Christopher)

The replica *Iron Duke* broad gauge loco, with 8-foot driving wheels, displayed at the Gloucestershire Warwickshire Railway at Teddington.

The functional central pier of the Royal Albert Bridge descends 70 feet beneath the water.

The Royal Albert Bridge as it is today and, behind it, the 1961 Tamar Bridge, which carries road traffic.

The Menheniot viaduct, with masonry extensions and steel girders which replaced the wooden supports.

The flying bridge at Liskeard Station springs direct from the slopes of the cutting.

The setting of the St Germans viaduct over the River Tiddy reveals the scale of the original. Brunel's timber structure consisted of fourteen trestles and was replaced by the present viaduct in 1907.

You can't get away from the chains at Millwall!

Recently uncovered, the hefty timbers of the *Great Eastern*'s slipway at Millwall on the Isle of Dogs.

A contemporary illustration of the *Great Eastern* in her prime. Note the uncluttered superstructure and the five funnels.

Isambard Kingdom Brunel, from a studio photograph.

CLIFTON SUSPENSION BRIDGE

LOCATION: Avon Gorge, Bristol, BS8 3PA. (OS map 172:565732)

GETTING THERE: Head for Clifton and the Suspension Bridge Road, or from the Leigh Woods side, take Bridge Road.

OPENING TIMES: The bridge is open twenty-four hours a day, but note that it sometimes closes when large events are held in nearby Ashton Court, such as the Balloon Fiesta in August. The visitor centre is open 10.00–17.00 every day, except 24–26 December and New Year.

Bristol's most spectacular landmark, day or night. Pedestrians and cyclists go free, while motorists pay £1. There are commemorative plaques galore, but no sphinx. Spot the differences between the two towers and look out for the plinth that once supported the iron cable strung across the gorge (beneath the parapet on the Leigh Woods side). Looking towards Bristol, Brunel's dock entrance is in front of the A3029 road swing bridge between the north lock seen on the left and the muddy Avon and Cut to the right.

INFORMATION: 0117 974 4664 for visitor services, or see www.cliftonbridge. org.uk

QUEEN SQUARE

LOCATION: Central Bristol, between the Centre and St Mary's Redcliffe.

GETTING THERE: From the Centre it is a short walk, or alternatively from Temple Meads head past St Mary's Redcliffe and across the old swing bridge over the Floating Harbour, then either right down Welsh Back or left along The Grove. There is limited parking on the square, plus awkward one-way streets.

Named after Queen Anne, this is the scene of the 1831 Bristol Riots. This eighteenth-century square was split in half by a busy through-road until 1999, but now the imposing equestrian statue of William III can be enjoyed in relative peace.

BRUNEL STATUE

LOCATION: Temple Quay, off the A4044.

GETTING THERE: A short walk from Temple Meads.

This bronze by John Doubleday, originally erected by the former Bristol & West Building Society in front of their tall monolithic office building on the Centre, is now at Temple Quay. Not my favourite Brunel statue as it has too much of the Charlie Chaplin about it for my taste.

Laying the first stone on the Leigh Woods side of the Clifton Suspension Bridge, 27 August 1836.

8

DESTINATION NEW YORK

Bristol and Mud!

Looking down from Clifton Suspension Bridge towards Bristol docks, you can actually see another of Brunel's bridges – his first using tubular iron girders. There are three openings from the river passing under the modern Brunel Way A3029 road. On the right, the Avon is channelled into the New Cut and next to it, in the middle, is the South Entrance Lock designed by IKB. His iron swing bridge, which spanned that lock, has been moved and now sits beside the newer lock on the left. Brunel worked on several dock improvement projects, including Sunderland's Monkwearmouth Dock, the Briton Ferry Dock at Neath in West Glamorgan, the Great Western Docks in Devonport and at Bristol. And while these solid ventures are often glossed over in favour of his more glamorous projects, they are rewarding hunting grounds in comprehending the bigger picture of an engineer working across a broad spread of disciplines.

In the early 1980s I had the good fortune to work in a former warehouse building on the quayside overlooking St Augustine's Reach. There was still some commercial activity within the docks then; Charles Hill & Sons building ships at the Albion Dockyard and the *Harry Brown* delivering its load of Bristol Channel sand and gravel to the sand wharf in Hotwells. It was a twilight period of gentle, almost romantic decay – the lull before a storm of redevelopment which saw the old docks born of the Industrial Revolution refashioned by the modern lifestyle

revolution into harbourside apartments and trendy bars. The docks had been the beating heart of this thriving city, which grew prosperous on the trade of tobacco, timber, sugar, cotton, spirits and the 'Africa trade' – that appalling human cargo of slaves. And yet, even at the dawn of the Industrial Revolution, Bristol's docks were in decline. In 1700 the port had been second only to London, by 1800 it was ninth on the list with other regional ports in ascendance.

The trouble lay with the tidal waters of the River Avon. Bristol is situated some eight miles or so from the sea and vessels had to negotiate its winding course to reach the docks, only then to suffer the ignominy of tidal waters that twice daily left them sitting high and dry on the mud banks; hence we have the expression 'ship-shape and Bristol fashion'. Something had to be done and, in 1804, work began on William Jessop's Floating Harbour, or 'Float'. This involved enclosing the river between the Neetham Dam at Temple Meads and Rownham Dam on the western side of the city to ensure that vessels remained afloat twenty-four hours a day. To accommodate ships either entering or leaving the Float, the Cumberland Basin was created, linking the docks and the River Avon. Vessels would navigate the river at high tide, enter the basin via entrance locks and were then 'locked in' before proceeding through a junction lock into the Float. Conversely, outgoing vessels congregated in the basin to await high water for the trip downriver. And to cope with the coming and going of the tidal waters, a wide channel called the New Cut skirted around the south side of the docks and

Andy King, former curator of the Bristol Industrial Museum, looks into the deep pools where the sluice gates are located at the Underfall Yard.

backwards and forwards across the harbour with a submerged blade scraping the mud towards the mouth of the Cumberland Basin, where tidal waters would shift it.

Bristol Docks: The Underfall Yard

The Dock Co. approved the conversion of the Rownham Dam into an underfall and also the construction of the drag boat, and this work was carried out between 1833 and 1834. Today the Rownham site is known as the Underfall Yard and it is one of the Bristol's better-kept secrets – a working shipyard that retains something of the atmosphere of the old docks. The Underfall Yard Restoration Trust conducts occasional tours, but in 2005 I was fortunate to have Andy King, then curator of the Bristol Industrial Museum, to show me around. We began at the water's edge, where the old dam sits unseen somewhere beneath our feet. From there an underground water track leads to sluices before heading out towards the road and the cut beyond. A few yards from the entrance gates to the yard an old iron sluice gate, about 7 feet high and looking like an enormous potato waffle, is propped against a red-brick building. Inside are four working sluice gates, which are hidden from view in deep pools of inky water. They are raised or lowered to adjust the flow of water on a daily basis, a process that is now automated and has been much modified over the years. As Andy explained, 'Brunel gets all the credit for the underfall, but it was a docks engineer named Griddleston who renewed most of it in the 1880s.'

linked up with the river again at Temple Meads, while in the centre of the city the River Frome topped up the float.

Jessop's Floating Harbour gave Bristol a second lease of life as a working port, but another problem soon became evident as mud and silt accumulated because the flow of water was now insufficient to wash it away. To tackle this, a stop gate was installed at the Prince's Street Bridge so that one half of the Float at a time could be isolated for cleaning; an arrangement which proved unpopular with ship operators. Something better was needed and through his new Bristol contacts, IKB was introduced to the Dock Co. in 1832 and they commissioned him to inspect the docks and report back with suggestions. To improve the flow of water to carry the mud particles away, IKB proposed raising the height of the Neetham Dam and installing sluice gates at the Rownham Dam so that the mud would be carried out into the cut. In effect this turned the 'overfall' dam at Rownham, acting like the overflow on a bath, into an 'underfall' and more like a plughole. That helped to stop mud accumulating, but the existing shoals of mud also needed to be shifted and Brunel designed a drag boat, which would be winched

The land where the Underfall Yard stands today was created by Jessop's original damming of the river. Most of the buildings and sheds date back to the second half of the nineteenth century while others, such as the harbourmaster's office, are later additions. You can see into some of the workshops, where traditional wooden boats are constructed or restored alongside modern fibreglass hulls. Over on the other side of the 1890 patent slipway, there is an imposing brick building with a square tower beneath an even taller chimney. This is the pump house, which to this day supplies hydraulic power for swing bridges throughout the docks via a system of buried iron pipes. Originally powered by steam, the engines have run on oil since 1905, and impressive though they are, it is the goings-on outside that caught my attention. A huge cylinder called a hydraulic accumulator is pumped upwards by the engines to the top of a steel shaft, before it descends by gravity to create hydraulic pressure when it is required. This extraordinary Heath Robinson contraption

The hydraulic accumulator at the Underfall Yard stores hydraulic power for the dock's many moving bridges.

was installed in 1954 to replace the original one inside the tower, and day after day it rides up and down the height of the building for all the world to see.

Back to Brunel and mud. Completed in 1834, his steam-powered dredger worked very successfully hauling itself on cables across the Float. When it was worn out, a second dredger – designated as the *BD6 for Bristol Docks 6* – was built to the same design and it continued in operation until 1961. In addition the *BD6* appears to have a close cousin, which was probably built in Bristol in 1844 to Brunel's original designs. Used to shift mud at Bridgwater, this boat was christened *Bertha* by a journalist and now resides at a museum in Eyemouth, Scotland.

The *Great Western* Steamship

Mud aside, Brunel had an ulterior motive in improving Bristol's docks as he envisioned it as part of an ambitious integrated transport system that was way ahead of its time. His decision to begin a shipbuilding career appears to stem from one of those wonderfully Brunellian

moments when a new idea is conjured out of the air and he says, 'Why not?' You can picture the scene as the board of directors of the GWR meet in a smoke-filled room in Blackfriars, London. One of them happens to comment on the inordinate length of the railway line. Brunel sucks slowly on his cigar, a curl of bluish smoke lingers in the air, and half in jest he responds. 'Why not make it longer, and have a steamboat go from Bristol to New York and call it the *Great Western*?' There is a moment of uneasy silence, glances are exchanged, and they return to the business of the evening. But one of the directors, Thomas Guppy, was prepared to take the idea seriously. Indeed, why not build a ship?

In January 1836 the Great Western Steamship Co. (GWSC) was formed with IKB as engineer – he was to give his services for free. The Bristol shipbuilder William Patterson was selected to build the ship at his yard near Prince Street Bridge, and in June 1836 the keel was laid. At 212 feet long and with a displacement of 2,300 tons, the *Great Western* was to be bigger than any ship afloat, but in most respects construction of the oak hull followed traditional methods, with ribs forming the bottom and sides. Great emphasis was placed on longitudinal strengthening, with extra stiffening provided by iron diagonals and a row of iron bolts running the full length.

To drive the paddle wheels, IKB called upon Henry Maudsley to build a 'pair of engines', so called because they had two cylinders, each of which could drive one paddle wheel or both, fired by four flue boilers. These would be fed with coal and once again IKB's old adversary Dr Lardner had something to say on the matter. At a meeting held in Bristol for the advancement of science, which IKB attended, he held forth on the prospects of a steamship being capable of crossing the Atlantic:

> As the project of making the voyage directly from New York to Liverpool, it was perfectly chimerical, and they might as well talk of making the voyage from New York to the Moon … 2,080 miles is the longest run a steamer could encounter – at the end of that distance she would require a relay of coals.

But what the good doctor failed to grasp was that whereas the capacity of the hull increases as the cube of its dimensions, its resistance only increases as a square of those dimensions. In other words, the surface resistance is proportionally smaller and hence so is the fuel requirement.

Not much is left to show where the *Great Western* steamship was built; a plaque on the dockside building and various models in museums.

The *Great Western* was launched on the morning of 19 July 1837, and contemporary paintings of the scene show her slipping into the water, cheered on by a crowd of 50,000 onlookers. She was the biggest vessel afloat and one observer recorded:

> … the beautiful and majestic vessel had glided into her adopted element, which she did like 'a thing of life' and floated gracefully and steadily on the water, the multitudes in every direction rent the air with their acclamations …

To fit the engines, it was decided to sail the ship round the south coast to the Thames and with her large paddle wheels carried as cargo, she was towed by steam tug out through the Cumberland Basin and up to the Bristol Channel.

When the first advertisements for transatlantic sailings appeared in Bristol newspapers in March 1838 they announced that the 128 state rooms were all of one class with a fare of thirty-five guineas, although twenty 'good bed places' had been allocated for servants, who travelled at half price. The ship would also carry some cargo, which was especially important on the return crossings, when she brought cotton for the Great Western Cotton Mill in Bristol. However, the company wasn't without rivals in the race to be first to offer regular Atlantic sailings. The

American Steam Navigation Co. in Liverpool aimed to steal Brunel's thunder by pressing the smaller *Sirius* into service and within a week their advertisements were offering three classes of travel. The race was on.

On 24 March, the *Great Western* began her first steam trials and despite a slight mishap – when her paddle wheels fouled a moored vessel – the ship handled well. Further trials followed and by the end of the month she was on her way to Bristol when a more serious incident occurred, as the official report recorded:

> At a quarter past 8 o'clock a strong smell of burning oil was perceived to arise from the felt cloth on the upper part of the boilers which soon afterwards took fire and from the quantity of dense smoke arising from it caused much apprehension.

The boiler lagging was on fire. In the ensuing panic, several of the stokers abandoned ship and began rowing to the shore. Christopher Caxton rushed into the smoke-filled boiler room to investigate as best he could when a heavy object fell upon him. Staggering to his feet, he discovered a body lying face down in a pool of water from the fire hoses. It was Brunel, who had fallen through a charred ladder rung. Once the fire was under control, Brunel was taken by boat to Hole Haven and although having been badly shaken and bruised, an examination revealed no serious injuries. Meanwhile the *Great Western* continued on to Bristol, but significantly not to the city docks because she could not fit through the entrance lock without removing her paddles. An alternative mooring was arranged at Kingroad, Pill, a few miles along the Avon. It was here that repairs were undertaken for her maiden voyage to New York. Meanwhile, on 4 April, the *Sirius* set off from Cork and was heading west. It would be another four days before the *Great Western* could follow.

Not surprisingly, the *Sirius* reached New York first; she had taken nineteen days, and according to legend had only managed to limp to the finishing line by burning furniture and fittings after her coal had run out. But that didn't matter to the ecstatic crowds, who swarmed to welcome her and the inauguration of a new era of transatlantic travel. It was a moment of great drama captured by one of the New York newspapers:

> The news of the arrival of the *Sirius* spread like wild fire through the city, and the river became literally dotted all over with boats conveying the curious to and from the stranger. There seemed

to be a universal voice of congratulation and every visage was illuminated with delight … suddenly, there was seen over Governor's Island, a dense black cloud of smoke spreading itself upwards. On it came with great rapidity …

It was the *Great Western*! Having left later than her rival and taken a longer route around Ireland, Brunel's ship had made up the time to arrive at full steam with paddles biting into the water. Lardner had been completely wrong with his calculations that it would take 1,348 tons of coal to make the crossing. The *Great Western* had only consumed 456 tons and she still had coal to spare.

While the transatlantic career of the *Sirius* was predictably short-lived – she had been built as a ferry between England and Ireland, after all – it was the *Great Western* that demonstrated the viability of regular scheduled crossings. During 1838 she completed five round voyages, with an average crossing time of just over sixteen days. She was soon joined by more capable rivals, the most significant coming in the form of Samuel Cunard's fleet of four vessels starting with the *Britannia*, which departed from Liverpool in July 1840. By this time the GWSC had started work on a second vessel, the revolutionary *Great Britain*, but it would be another five years before her first transatlantic foray, and by then the lucrative mail contract had been awarded to Cunard.

The *Great Western* was a fine ship and established an exemplary record for both reliability and speed, but IKB and his company directors were feeling increasingly hampered by the intransigence of Bristol's Dock Co. In particular it had failed to address the issue of increasing ship sizes, especially the new steamships – a problem spurred on by IKB himself with his desire to super-size. As early as 1835, Brunel had warned them that the existing 33-foot-wide South Entrance Lock into Cumberland Basin was too narrow but, strapped for cash, they only gave the go-ahead for improvements in 1845, further exacerbating Bristol's decline. IKB realised that the writing was already on the wall for the city as a great port and, given that his attention was on other projects, his enthusiasm for improving the lock had waned, as demonstrated in this comment to the dock company:

I have recommended these dimensions because I believe they would be sufficient to accommodate all ordinary Steam Boats built for the Irish Trade – this I now believe is sufficient for the Port of Bristol.

Ordinary steamboats and the Irish trade? What had become of the transatlantic connection he had so much coveted a decade earlier? All gone! The historian and Brunel biographer Angus Buchanan has summed up this statement as 'an epitaph on the Floating Harbour'.

Bristol Docks: South Entrance Lock

The new South Entrance Lock was eventually completed in 1849 and measured 262 feet by 54 feet featuring elegantly curved walls for maximum clearance of laden vessels. To seal the lock, IKB devised a system of iron 'caisson' gates divided into three chambers, one of which contained a pocket of air to create partial buoyancy in high water. They were moved on cast-iron rails via a manually operated capstan, hinging sideways into recesses within the dockside masonry, although apparently this semi-buoyancy had a tendency to tear the gates from their hinges at high tide. To allow road traffic to pass over the lock IKB came up with another innovation, a swing bridge constructed with sides formed by tubular wrought-iron girders; the direct predecessor to the iron girder bridges at Chepstow and Saltash.

To carry road traffic over his South Entrance Lock, Brunel built this swing bridge with circular iron girders – a direct precursor to the Chepstow and Saltash bridges. Confusingly, it is now beside the later North or New Entrance Lock (not Brunel's work), and a replica sits over his lock.

Brunel's South Entrance Lock did not remain in operation for very long and in 1873 the docks engineer Thomas Howard constructed the larger New Entrance Lock just to the north of it. To save money, Howard cheekily pinched Brunel's swing bridge for his lock because he considered the old lock to be redundant. When the boat operators complained, the corporation agreed to keep the south entrance operating and so Howard dusted off IKB's drawings and built an identical bridge. Unfortunately the Brunel Lock, as it is generally known, fell into disuse because mud tended to accumulate under the gates and by the 1890s a steamer landing grid had been constructed, blocking access from the river. The lock is still there and is best viewed at low tide to see the curved bottom and recesses for the gates, which were removed in 1906. IKB's original tubular girder bridge sits parallel to the dockside beside the new lock in the shadow of the modern Plimsoll swing bridge. There have been calls for it to be swung back over the lock to complete a round-the-docks cycle path, but the practicalities of manually operating an additional bridge make this unlikely for the foreseeable future.

The completion of the South Entrance Lock had come too late to help the GWSC and they continued to operate the *Great Western* from the Kingroad moorings with incoming passengers transferred to smaller boats, known as lighters, to complete their journey. By 1842 she was using Liverpool to avoid Bristol's exorbitant harbour dues. This marked the end of any hope for a transatlantic transport system from London via the GWR and Bristol. Following the *Great Britain*'s grounding at Dundrum Bay in 1846 (*see* Chapter 9), the steamship company was left in severe financial straits and in December 1846 the *Great Western* was taken out of service. In eight years of operation she had completed forty-five transatlantic crossings, carrying almost 8,000 passengers. She was laid up in Bristol and offered for auction in 1847 but, when the bidding only reached £20,000, was withdrawn from sale. The West India Royal Mail Steam Packet Co. came up with a better offer and for the next ten years she continued to traverse the Atlantic with only a short break in 1855 to take troops to the Crimea. In 1856 she was sold to ship-breakers on the Thames and Brunel's first ship was broken up at Millbank just as his last, the *Great Eastern*, was taking shape on the Isle of Dogs.

The *Great Western* is often relegated to second billing in comparison with IKB's later ships. It was never photographed, nor does it survive. In Bristol Docks there is a modest plaque indicating where Patterson's shipyard once was, erected in 1938 to mark the 100th anniversary of her launch. The building on this site was bombed out during the Second World War and the plaque has been remounted on what was part of the old Bristol Industrial Museum on Prince's Wharf, which is now known as the M Shed. Inside the new museum, there is little to commemorate Brunel's contribution to the city. A little further along the dockside, the SS *Great Britain*'s museum houses several items, including the ship's brass bell. And that's it. What a great disservice to what was Brunel's first great ship and the direct forerunner of the other two.

Great Western Hotel

In St George's Road, tucked behind the curve of Bristol's Council House, there is an elegant colonnaded building known as Brunel House. Built between 1837 and 1839, this is the former Great Western Hotel designed by R.S. Pope in collaboration with IKB to accommodate the transatlantic travellers overnight between train and ship. It ceased to be a hotel in 1855 and for a time functioned as a Turkish baths, and currently serves as offices. Only the facade remains from the original; you can still see the archway at either side where carriages entered and left the rear courtyard, much as they did at Temple Meads. It is a tangible relic of Brunel's transatlantic dream.

The site of the Great Western Hotel on St George's Road, Bristol.

UNDERFALL YARD AND SOUTH ENTRANCE LOCK

LOCATION: Yard is off the Cumberland Road, BS1 6XG. The South Entrance Lock is beneath the A3029 Plimsoll Bridge over Cumberland Basin.

GETTING THERE: It is a reasonably long walk from the *Great Britain*, but it is usually easy to find somewhere to park at the Underfall Yard or the Cumberland Basin area.

Underfall Yard gives a taste of how the working docks once felt. The Underfall Yard Restoration Trust conducts tours.

Brunel's South Entrance Lock into the Cumberland Basin with its curved bottom is still visible, although the lock gates have been replaced by a concrete wall. The little bridge over the lock is a copy; the Brunel original sits beside the nearby North Entrance Lock.

INFORMATION: www.underfallboatyard.co.uk

GREAT WESTERN LAUNCH SITE

LOCATION: Princes Wharf, off Wapping Road.

GETTING THERE: Aim for the southern end of Prince Street Bridge by the M Shed Museum. Look for the tall dock cranes.

Spot the circular plaque high on the wall.

GREAT WESTERN HOTEL

LOCATION: St George's Road, near Bristol's city centre (shown opposite).

GETTING THERE: The hotel sits behind the Council House building, just off Park Street. Some parking.

It was designed by R.S. Pope in collaboration with Brunel. Now known as Brunel House, it was constructed to accommodate passengers from Paddington before catching their boat to America. Only the facade remains as part of an office building – note the arches either side for the carriages to enter and exit the hotel. There is a plaque on the left-hand side.

M SHED

LOCATION: Princes Wharf, Wapping Road, BS1 4RN.

OPENING TIMES: Tuesday–Wednesday, 10.00-17.00. Saturday/Sunday and Bank Holidays 10.00–18.00. Admission free.

Although the *Great Western* was built here, the museum situated within the old dockside transit sheds has little on IKB.

INFORMATION: 0117 352 6600 / www.bristolmuseums.org.uk/m-shed

BERTHA

OPENING TIMES: April to November, 10.00–17.00 daily.

This little drag boat built for the Bridgewater Canal now sits in pride of place beside the World of Boats Museum at the Eyemouth Maritime Centre in Scotland.

INFORMATION: www.worldofboats.org

The former Great Western Hotel in St George's Road was built to accomodate the transatlantic passengers travelling through Bristol.

9

THE SS *GREAT BRITAIN*

Bristol

Originally Brunel had envisioned a second paddle-driven wooden-hulled ship to follow in the wake of the *Great Western*'s transatlantic successes. The African oak had been purchased and the ship even had a name: *The City of New York*. He knew that iron is both cheaper and stronger than wood, but its use for ocean-going vessels was restricted because working the iron created a magnetic field, which would throw out the ship's compass. This problem was overcome by a system of correcting magnets devised by the Astronomer Royal and when the iron-hulled paddle steamer *Rainbow* arrived at Bristol in 1838, equipped with the apparatus, Brunel readily swapped to iron for his new ship. When the experimental *Archimedes* came to town in the following year, pushed along by an experimental screw propeller designed by Francis Pettit Smith, he did away with the paddles. It was with the resulting combination of an iron hull with a screw propeller that Brunel defined the shape of modern shipping.

On 19 July 1843, Prince Albert travelled to Bristol by train and, along with 520 guests (including Marc Brunel) plus thousands of onlookers, he watched as Mrs Miles, the wife of one of the GWSC directors and who, incidentally, had also launched the *Great Western*, stepped forward to name the ship as the *Great Britain*. When Mrs Miles swung the bottle of champagne it missed its mark, so the prince grabbed another bottle and hurled it against the bows, showering bubbly and glass upon the workmen down below. The ship remained in Bristol to be fitted out and was not ready for sea trials until late the following year. On the first

The hulk was towed back from the Falkland Islands, riding piggyback on a huge pontoon. (Campbell McCutcheon)

Taken shortly after the ship's return to Bristol, this photograph shows the extent of the rust and the scale of the task facing the restorers.

and picking my way between the rusty bulkheads evoked a real sense that this was the real *Great Britain*. But it couldn't last. No, literally. If they had left it like that the ship would probably have crumbled under its own weight into a heap of rust. And even though I approached the revamped *Great Britain* with some trepidation, I needn't have worried because the result of all that money and hard work is absolutely superb!

A Dramatic Life

A new museum has been created in the old dockside buildings, its balconied interior dominated by the huge iron main yard, 95 feet long and weighing over 17 tons, stretching from one corner to the other. Running around it, a series of displays leads visitors through the story of the ship and its restoration, highlighting the many chapters in its long career.

Despite having the capacity for 252 passengers, when the *Great Britain* made her first transatlantic run to New York in 1845 she sailed with just fifty on board because of fears about the safety of the novel construction methods. But this ship was a beauty and, under the command of Lieutenant James Hoskin, who had commanded the *Great Western* on thirty-two voyages, she handled well and was said to cut through the water like a sloop. She arrived safely in New York after fourteen days and eleven hours and at a very respectable average speed of 9.4 knots.

Passenger confidence picked up on subsequent sailings and by the fifth voyage in September 1846, she sailed with 180 passengers and a considerable amount of cargo. Unfortunately Hoskin's skills as a navigator did not improve. He had already scraped her bottom on the Nantucket shoals on the second voyage, badly damaging the propeller. Now he almost destroyed the ship when he confused a newly commissioned lighthouse on the Irish coast for the Chicken Rock Light on the southern tip of the Isle of Man. In the darkness and driving rain the *Great Britain* ran aground at Dundrum Bay, County Down, and with the dawn came the realisation of how close the ship had come to being thrown against rocks. Although leaking slightly, the hull was in good shape, but there was no immediate hope of refloating the ship. When IKB arrived at the scene to examine the measures taken to protect the *Great Britain*, he was appalled by what he found:

attempt to get her through the Cumberland Basin she became stuck because the Dock Co. had not widened the locks as expected by that time. Some critics suggested she would never leave Bristol, but on the second attempt IKB had some of the dockside masonry removed and the 322-feet-long hull literally scraped through.

On my own tour of IKB's work I had left my visit to the SS *Great Britain* until last because a major £11.3 million makeover for both the ship and historic dock had been underway. My interest in the ship goes back twenty-five years to the early 1980s, when it had been little more than a rusty hulk. At that time, Baltic pine cladding – added during its later life as a cargo ship – was heaped on one side of the dock with nothing stopping visitors helping themselves to a piece. On board there was a token area of weather decking and the whole vessel was stripped to its bare iron bones with only your imagination and some outlines painted on wooden panels to indicate the different areas. And I liked it like that. I have always been wary of the trend to over-restore relics,

I was grieved to see this fine ship lying unprotected, deserted and abandoned by those who ought to know her value … the finest ship in the world, in excellent condition, such that four or five thousand pounds would repair all damage done, has been left and is lying like a useless saucepan kicking about on the most exposed shore that you can imagine.

To shield her from the winter storms, he devised a barrier of wooden faggots reinforced with iron rods and chains. In the spring her plates were repaired and finally, in August, she was pulled clear by HMS *Birkenhead*. Few other ships would have survived such punishment, but the GWSC could not survive the costs of saving her. Both the *Great Western* and *Great Britain* were sold, the latter in 1850 for a fifth of her original cost. Bought by Gibbs, Bright and Co. of Liverpool, she was refitted for the 12,000-mile run to Australia, where the discovery of gold was drawing immigrants by the thousands. To reduce the demand on coal for such a long voyage, she became a sailing ship principally with a capability to steam if required. The number of masts was reduced to four, although the sail area was increased, and twin funnels were installed for a new two-cylinder oscillating engine. Passenger capacity was increased to 700 by adding an upper deck and converting the saloon deck into further berths and cargo space.

Between 1852 and 1875, the *Great Britain* made thirty-two voyages to Australia, with a ten-month gap in 1854 when she carried troops to the Crimea. When she became too old for the Australia run, the engines were removed completely and she was converted into a cargo-carrying windjammer in 1882. It is said that in a good wind, the ship could make speeds as fast as many a clipper. Then in 1886 she ran into trouble rounding Cape Horn and her captain sought refuge in Port Stanley on the Falkland Islands. Too expensive to repair, she was sold for £2,000 as a floating store for wool and later coal. By the 1930s she wasn't even up to that job and she was she was beached and abandoned at Sparrow Cove.

Walking around the ship today, you can only marvel at the scale of the salvage operation that brought her back to Bristol in one piece. In 1967 a naval architect, Dr Ewan Corlett, had written to *The Times* urging the ship to be recovered, or at the very least, fully documented. A survey revealed that she was in remarkably good shape, despite a 13-inch crack caused by the constant action of the sea and, when the businessman Jack Hayward came forward with funding, a rescue operation was mounted. Because the ship was no longer seaworthy, a direct tow was out of the question and the Hamburg salvage company of Ulrich Harms came up with a submersible pontoon to carry her back in 1970. At Avonmouth the pontoon was removed and on 19 July – the anniversary of her launch – she rode on the spring tide back into the Great Western Dock.

The Trouble with Rust

The first thing you see when entering the dockyard nowadays is the ship's elaborately decorated stern, featuring the gilded arms of the city of Bristol. The next thing is the new 'glass sea', which seals the hull within its own controlled environment beneath 169 panes of plate glass and 2 inches of water. Looking down into this fish tank from above, you can make out shapes moving around and you suddenly realise that these are undersea people. Before exploring the rest of the ship itself I had to go down there, and access is via a new staircase or lift – they have done wonders making all areas accessible to everyone with wheelchair lifts and ramps. Beneath the glass it is strange to see the dry dock so, well, dry. This is architecture meeting museum head-on; a ship in a bottle, almost. A huge dehumidification machine known as 'Deep Thought II' – its companion is hidden within the hull – creates an atmosphere said to be as dry as the Arizona desert. Huge steel air ducts then carry the warm, dry air the length of the hull and gently blow against the iron plates.

The hull is clinker-built, constructed of iron plates forged at the foundry at Coalbrookdale and shipped down the Severn. Each plate is about 6 feet by 2.5 feet, individually shaped and riveted in position, overlapping horizontally and connected on the inside by iron straps. The result is a remarkably strong hull, and a very graceful one too seen from underneath; note the bulge, which increases internal space. The great enemy of iron is, of course, rust. Rust is nasty stuff. You can't just flake it off and paint over the bare metal. A lifetime of sitting in seawater has left the hull impregnated with salts, which have bonded with the metal and attract water to continue the corrosive process. Only by keeping the humidity to 20 per cent or less can the process be arrested.

One of the first dilemmas tackled by the trust that cares for the *Great Britain* was deciding on which period of her life to base the reconstruction work. Six, five or four masts? Transatlantic liner, emigrant ship or windjammer? In the end they settled on the original configuration as befits her return to Bristol and one obvious sign of this is at the rear of the hull, where the huge six-bladed screw propeller and rudder are located. IKB devised a balanced rudder with its weight distributed on either side of the vertical iron column called the rudder stock. This system meant that the crew could turn the rudder with much less effort than a conventional one hinged at the forward end. An example of the later 1857 two-bladed screw can be seen in the museum and visitors can raise or lower it within the stern frame to reduce drag while under sail.

Back to the ship above water. The deck is surprisingly bare – ideal for operating the sails. Skylights allow light into the saloons and a glazed roof reveals the engine space where passengers could watch the great drive-wheel at work. A little behind this, there is a white line on the decking and a sign saying, 'First class passengers only beyond this line.' Crossing the ship just behind the 38-foot-high funnel is the bridge, which is completely open to the elements.

Several stairways lead down inside the ship. Firstly there is the promenade deck, a surprisingly big space where the saloon (first-class) passengers could stroll and socialise. Contemporary commentators fell over each other to sing the praises of the ship's interior, as this report from June 1845 reveals:

> The cabins are superb. The principal saloon is 110 feet long, the fore saloon is 65 feet, and both are upwards of 8 feet high. Beneath them are dining saloons of proportionate dimensions.

A reconstruction of the ship's original six-bladed propeller.

All these apartments are tastefully embellished … In a word, the character of the ship is sunk in the beauties and lordly air of a noble mansion.

Nowadays the promenade is populated by clusters of tourists, who mill about with headphones clamped to their heads, speaking too loudly because they can't hear themselves. Visitors are offered a choice of 'audio travel companions', including a first- or third-class passenger, a maritime archaeologist or even Sinbad, the ship's cat. It is a great system, but for those who prefer not to be plugged in, there are precious few information signs. On either side of the promenade several doors lead to cabins, some of which have been dressed with tableaux or scenes such as the ship's surgeon at work. Note his list of crew on sick parade. These cabins seem tiny, the inner ones are without windows and in one scene a woman passenger is being sick while a rat sits on the bedding. First-class ladies at least had their own boudoirs:

> About twelve sleeping berths on each side of the deck will be reserved for ladies, as they are made to communicate with two commodious ladies' boudoirs or private sitting rooms, measuring 17 feet by 14 feet. The advantages of this arrangement must be obvious, as ladies who are indisposed, or in negligé will be able to reach their sleeping berths without the slightest necessity for their appearing in public.

Directly beneath the promenade is the equally spacious first-class dining saloon and pools of light filter down through the skylights. Large mirrors at one end accentuate the space, which could accommodate 360 diners on rows of benches stuffed with hair. This contemporary description from the *New York Daily Tribune*:

> Down the centre are twelve principal columns of white and gold, with ornamental capitals of great beauty. Twelve similar columns also range down the walls on either side. Between these latter and the entrances to the sleeping berths are, on either side of the deck, eight palisters, in the Arabesque style (of which character the saloon generally partakes) beautifully painted with oriental birds and flowers … The archways of the several doors are tastefully carved and gilded, and are surmounted with neat medallion heads. The walls of this

On 15 April 1845, huge crowds turned out to see Queen Victoria and Prince Albert visit the *Great Britain* at Blackwall on the Thames, where it had been fitted out. The queen had admired the enormous length of the ship.

No longer seaworthy, the *Great Britain* is shown being mounted on the pontoon for the journey back to Bristol. Note the wooden cladding that covered the hull. (SWNA/JC)

apartment are of a delicate lemon tinted drab hue, relieved with blue white and gold.

At the centre of the ship is the vast engine bay, reaching from the keel right up to the weather deck. It has been fitted with a replica of Brunel's original engine, which consisted of four large inclined cylinders. A contemporary newspaper described its workings:

The bewildering mass of machinery which meets the view … has to perform one very simple duty, which is to turn the screw … at the stern with sufficient velocity to propel the ship. To effect this, four engines the united power of which equals that of 1,000 horses are employed. Their action upon the wheel is readily understood. There is an enormous wheel or drum, 18ft in diameter, working on an axis or spindle. To either end of the spindle is attached an enormous crank moved by one pair of engines the other pair driving the crank at the opposite end of the axis, so that the whole four expend their force upon the gigantic drum to whirl it around. The duty of this large wheel is to cause a band composed of four iron chains to revolve with it that in the regions below they may pass around and turn another and smaller wheel. Though weighing seven tons the chains work quite silently. The little wheel below has for its axis one end of a long horizontal shaft to the other extremity of which is attached the propeller.

Manufacturing large components such as the propeller shaft brought about the development of new precision tools. Until then the size of metal components had been limited by the capabilities of the existing tilt-hammers, but Nasmyth's mighty steam-hammer changed all that.

Ahead of the engine bay is more accommodation including the galley and food stores for fresh meat or fowl, packaged foods such as flour and grains and tinned soups, veal, jam, salmon and vegetables. The galley was fitted out with 'every conceivable apparatus for roasting, boiling, frying, grilling and stewing'. Steerage passengers made do with porridge, soup, ship's biscuits and perhaps a little salted pork. Beyond the galley are the steerage berths: basic wooden bunks, four in a tiny cabin or placed in long rows. It is said that steerage passengers were 'packed as close as bees in a hive'. Clothes are strung up between the bunks, in one cabin a woman is giving birth, and in another a cricket bat mounted on the

Left: The *Great Britain* rides the spring tide for her final journey to her dock in Bristol on 19 July 1970. Thousands of onlookers lined the Avon Gorge to see the ship pass under Clifton Bridge for the first and only time. (SWNA/JC)

Below: This photograph, taken looking forward in the mid-1970s, shows the replica skylight in position on the deck. (SWNA/JC)

Back in the Great Western Dock, where she was built, work begins on cleaning the wrought-iron hull. (SWNA/JC)

wall is a reminder that in 1861 the All England cricket team travelled on the ship for the first Australian tour.

Towards the bow, the area under the forecastle remains unrestored. This would have been the accommodation for the officers and, down below, for the crew. Today it reverberates with the steady thrum of the dehumidifier and in places daylight shines through holes in the iron plate. It is a reminder of what this ship was like before restoration and I hope they leave this one area as it is. As for the rest of this lovely ship and its dockside setting, it is a triumph … apart, perhaps, from the plastic cow! Go and see.

SS *GREAT BRITAIN*

LOCATION: Great Western Dock, Gas Ferry Road, BS1 6TY.

GETTING THERE: The ship is off the Cumberland Road, on the north side of the River Avon. Plenty of parking. Or walk along the dockside – *see* Round Trips on pages 143–5.

OPENING TIMES: Daily 10.00–17.30 in summer, or 10.00–16.30 winter (not 24, 25 December or second Monday in January).

The iron ship fits the dry dock like a hand in a glove. A dockside building houses an impressive exhibition with many artefacts, including some from the *Great Western* and *Great Eastern*. The glass sea keeps the hull and the visitors dry. At times the *Matthew* is moored nearby – a replica of the wooden ship John Cabot sailed to Newfoundland in 1497.

INFORMATION: 0117 926 0680 / www.ssgreatbritain.org

THE BRUNEL INSTITUTE

LOCATION: Set alongside the *Great Britain* in the Great Western Dockyard, this houses an extensive collection of books and items relating to the life and work of IKB.

OPENING TIMES: Closed Mondays. Tuesdays, Wednesdays, Fridays 10.30–16.30. Thursday 10.30–17.30. Open first two Saturdays of each month 10.30–16.30. Closed Sundays. Admission is free.

INFORMATION: 0117 926 0680

10

EXTENDING THE BROAD GAUGE

On Parallel Lines

The myth of Isambard Kingdom Brunel as the 'Little Giant' who packed some big ideas is an enduring one. Small but perfectly informed? Well no, not always. Never one to do something the way others had done it before, his groundbreaking solutions sometimes proved unworkable given the circumstances of the times. This was certainly the case with the broad gauge. The concept behind the broad gauge was fine, and once IKB had sorted out the initial problem with the see-saw motion, it gave a far superior ride to the narrower gauge favoured by Stephenson. But the conflict came where one gauge met the other and by 1845 there were ten such places. Gloucester station, for example, became notorious for the constant chaos of passengers changing from one gauge to another, and it is estimated that goods traffic could be delayed for up to five hours. IKB's solution to this 'break-of-gauge' problem was the transfer shed, where both broad and standard gauge lines came together separated by a narrow platform. Passengers would, he argued, 'merely step from one carriage into the other and on the same platform'. There is a fine example of a transfer shed at Didcot, Berkshire. This had been a sleepy little village before the railway came, but its location made it strategically important for the expansion of the railway as the junction for a line to Oxford and, so Brunel hoped, to the industrial north. Today it is home to the Great Western Railway Society, one of the biggest preserved steam centres in the country. Most of the collection is from

the golden age of the GWR in the first half of the twentieth century, with lots of big green locos and carriages in distinctive chocolate and cream liveries. But head past these and you reach the broad gauge area and the long wood-clad transfer shed. Inside, the central platform separates the two tracks by only a few yards and the gib of a crane indicates how the luggage and goods would have been moved from one train to the other.

In 1845, a royal commission was appointed to examine the gauge issue and hear evidence from the pro- and anti-broad gauge lobbies, although admittedly mostly from the latter. However, while the commissioners seemed to appreciate the advantages of the broader gauge, they could not ignore the fact that it accounted for only 274 miles of line compared with 1,901 of narrower gauge. IKB may have cornered the railway engineering market in the south and south-west – partly through the ripple effect of personal contacts and partly for pragmatic reasons of shared line and facilities – but his 7-feet-and-a-bit gauge had not won many converts elsewhere. In the end it was simply a case of being out-gunned, and a good idea gave way to a lesser one because of the weight of numbers. Inevitably the commission recommended that 'uniformity be produced by an alteration of the broad to narrow gauge'.

In the following year the Act for the Regulating of the Gauge of Railways specified 4 feet 8.5 inches as the gauge for all new railways, although a clause did allow for a couple of already planned broad gauge

Broad Gauge Survivors

The quadrupling of the mainline over the years, combined with the reduction in stopping trains, has resulted in the loss of many of the original smaller stations between Paddington and Bristol. Fortunately on the quieter branch lines, many examples of buildings from the broad gauge period have survived. One of the finest is to be found at Culham station, on the Oxford line a few miles north of Didcot. Brunel's designs for the GWR stations came in a series of standardised sizes and layouts and this is a typical 'Pangbourne' type of second-class or minor station. Bourne's description of Pangbourne serves Culham perfectly:

> The station, as is usual with those of this class, is composed of a house, placed on the side of the railway most accessible to the public … a covered platform being placed on the opposite side. The general style is Elizabethan. It consists of one storey only, and is divided into a booking office with a bow looking upon the railway, and two waiting rooms. The eaves of the building are produced so as to form a complete covering all round the house, and extending over the platform …

When broad and standard gauge met head-on, the result was chaos, as shown in this engraving of the scene at Gloucester, where passengers had to swap trains.

connecting lines to be completed. In some cases, mixed gauge track was created by adding a third rail and there is a reconstructed section of this at Didcot which illustrates the additional complexity and materials this entailed, especially at the points. Parliament had sounded the death knell for the broad gauge, although a look at railway maps of the 1860s reveals how extensive the network had become at its pinnacle. There was the mainline from London to Bristol of course, then south-west down through Exeter and Plymouth to Penzance, southwards to Salisbury and Weymouth, north from Didcot to Oxford, and via Cirencester and Stroud to Gloucester, leading to Hereford or across South Wales all the way to Milford Haven in Pembrokeshire.

The tracks might be narrower nowadays, but the old GWR mainline still serves as the greatest monument to the broad gauge era. The shallow gradients of Brunel's Billiard Table and his wide tunnels and cuttings have ensured that today's high speed trains can travel at 125mph. Imagine what might have been possible with a modern train had the 7-feet gauge won through.

Lovingly restored, Culham station now serves as a commercial office, so you can't get inside, but the outside is still worth the trip. The Railway Inn on the other side of the track is reached by a new footbridge built about four years ago. 'My husband bought the old one!' the landlady told me, and with that she showed me into the garden to see it. Constructed of metal girders with latticework in-fills, it is not a Brunellian design and probably dates from the 1930s.

To find another broad gauge survivor, I headed down to Frome station on the former Wilts, Somerset & Weymouth Railway, which opened in 1850. Hidden amid a scattering of industrial buildings, this is probably the worst signposted station in Britain. I call it 'the station that time forgot'. I could tell at a glance that it is not over-used – the handful of parking spaces allocated to passengers for Wessex Trains were only half occupied. The booking office was closed, predictably, and a nicotine-stained sign pointed to the side gate for access. The wooden building is a pleasing structure designed by one of IKB's many assistants, J.B. Hannerford, and features an all-over roof. A single track runs through now and as the far platform has been cut off, it is slowly succumbing to weeds. Otherwise it has been restored and was

Pretty in pink, the station at Frome with its all-over wooden roof was designed by IKB's assistant J.B. Hannerford. But what happened to the train shed?

painted a somewhat uniform salmon pink colour fairly recently. And yet something important was missing. Where was the wooden train shed? I consulted a photograph of the shed published about twenty years ago, and damn fine it looked too. But in the place where the shed should be, there is … nothing! So if you think that in this day and age such relics are safe from destruction, then think again.

Another shed that could easily have shared this fate a few years ago is now protected as a listed building. The Cheltenham & Great Western Union Railway from Swindon via Cirencester across to Gloucester arrived in Stroud in April 1845. IKB was the engineer, although previously he had confessed to being unenthusiastic about the project:

Cheltenham Railway … Do not feel much interested in this. None of the parties are my friends. I hold it only because they can't do without me – it's an awkward line and estimates too low. However it's all in the way of business and it's a proud thing to monopolise all the West as I do.

Despite such misgivings, Brunel managed a very workmanlike job, as the buildings at Stroud demonstrate. It is a charming little station with Paddington/Swindon trains departing on the north platform and Stonehouse and Gloucester trains on the other. A footbridge was added after accidents involving passengers crossing the lines. While the station building is a nice example of the Brunel style, but nothing exceptional, the real gem is the old goods shed beyond the end of the platform. Built to broad gauge proportions, this is a solid and roomy stone building with sturdy wooden rafters overhead. The twinned arches at both ends lead to an internal platform for easy loading of goods onto carriages. At the

The transfer shed at Didcot – broad gauge on one side, standard on the other.

The ghost of Brunel – an 1892 cartoon from *Punch* to mark the passing of the broad gauge.

The marvellous twin-arched and stone-built goods shed at Stroud, complete with original advertising slogans painted on the trackside wall.

station end a narrow two-storey building provided office space. Note the old GWR advertisement painted along the trackside wall of the shed. 'GWR STROUD STATION -- EXPRESS GOODS TRAIN SERVICES – ONE DAY TRANSITS BETWEEN IMPORTANT TOWNS.' The trains no longer stop at the shed, and it is being reused a community arts centre.

The broad gauge had been a marvellous experiment that cost the GWR dearly and the directors had no choice in the end but to accept the inevitable. Thirty-three years after Brunel's death, the remaining track was converted to standard gauge in a mammoth operation involving thousands of men over one weekend in May 1892. *Punch* marked the 'Burial of the Broad Gauge' with a full-page illustration of the ghost of Brunel walking beside its grave, and in verse:

Lightly they'll talk of him now he is gone,
For the cheap 'Narrow Gauge' has out-stayed him,
Yet BULL might have found, had he let it go on,
That BRUNEL's Big Idea would have paid him!

But the battle is ended, our task is done;
After forty years' fight he's retiring.
This hour see thy triumph, O STEPHENSON;
Old 'Broad Gauge' no more will need firing.

South Devon: The Atmospheric Caper

Another of Brunel's transport innovations proved expensive to investors on the South Devon coast where he built a railway with atmosphere. Real atmosphere.

The Parliamentary Act for the construction of a railway connecting Exeter and Plymouth was passed in July 1844. In essence, this was a continuation of the Bristol & Exeter Railway's broad gauge line and not surprisingly the B&ER's engineer, IKB, was also appointed to the South Devon Railway. The chosen route followed the estuary of the River Exe, hugging the coastline through Starcross, Dawlish and Teignmouth and along the River Teign to Newton Abbott where it headed across country via Ivybridge to Plymouth. The result was one of the most picturesque stretches of railway imaginable, but this difficult undulating terrain with its many sharp corners and steep gradients was not well suited to the operation of the existing steam locomotive technology. This led IKB to adopt a radical alternative already arousing considerable interest in engineering circles; harnessing the atmosphere to generate motive power. Such a scheme had been proposed for the London and Croydon line, and a working atmospheric system had been installed between Kingstown and Dalkey in Ireland.

The principle of the atmospheric railway is simple, whereas the practicalities are not. The trains ran on conventional rails – broad gauge for the South Devon Railway inevitably – with an iron pipe with a continuous slot at the top laid between them. A piston moving inside the pipe was connected through a slot to a special carriage at the front of the train. Stationary steam engines pumped air out of the pipe ahead of the piston and the partial vacuum sucked it along the pipe.

Pumping stations were constructed along the South Devon line at Exeter, Countess Weir, Turf, Starcross, Dawlish, Summer House and Newton Abbott. When it opened in 1847, all was well. Speeds of 68mph were recorded for light trains and 35mph for heavier trains. The passengers appreciated the smooth run and the absence of smoke and smuts. But the devil is in the detail and niggling technical problems began to emerge. The movable leather flaps that sealed the slots began to let air and water leak in, and a poor seal meant a weak vacuum, resulting in increased coal consumption for the overworked pumping engines. During the summer months, the leather became hard and inflexible and when grease was smeared on it to prevent this from happening, the local rat population found the taste to their liking. In winter the water-sodden leather froze hard. Critics have also pointed to the impending problem of pipes crossing over at junctions, although there is no evidence that Brunel intended extending the atmospheric system beyond this particular line. After eight months, IKB was forced to admit that the system was unworkable and from September 1848 ordinary steam locomotives were brought in to work the line.

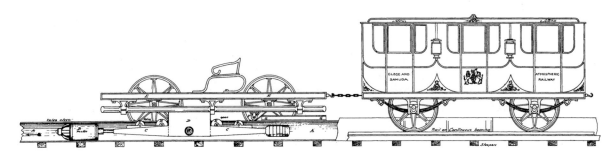

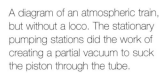

A diagram of an atmospheric train, but without a loco. The stationary pumping stations did the work of creating a partial vacuum to suck the piston through the tube.

Left: One of the largest sections of atmospheric pipe that you will find on display is at Didcot. The diameter of pipe varied depending on the steepness of any inclines. Leather flaps sealed the slot at the top. (Ute Christopher)

Below: Its chimney somewhat reduced, this former atmospheric railway pumping station at Starcross is the best surviving example. A museum for a while, it now serves as a boathouse.

The best surviving example of this 'atmospheric caper', as it became known, is the pumping house at Starcross, sandwiched between the road and railway line. Driving into the village, you can't miss the stumpy reddish chimney, which looks more like a church tower in the distance. Built in IKB's Italianate style, it appears to be in reasonably good repair, although the ornate chimney top has gone. There was a little Atmospheric Railway Museum housed here at one time, but it is now used as a boat house. Other pumping stations, at Torbay and Totnes, have been incorporated within commercial premises. If you want to see some of the atmospheric pipe, the best piece on display is at Didcot.

Dawlish, Newton Abbott and Brunel's Toilet!

Beyond Starcross, the railway tracks ride on the buttressed sea wall hugging the brick-red cliffs; one of the most wonderful backdrops you might wish for and it remains an irresistible draw to train enthusiasts. Dawlish comes next, a typical small seaside town that has seen better days and is not helped by the railway cutting it off from the beach. While I am a fan of IKB, I have to say that in the light of modern planning laws, this line passing along the seafront is a terrible eyesore. It is interesting to note that when Brunel bought land for his retirement home near Teignmouth (subsequently never built), he was the first to object at the prospect of new gasometers spoiling his view. Dawlish station is a twentieth-century building that stinks of urine and is devoid of any architectural merit – or a footbridge to the beach side for that matter. Its only saving grace is a bijou signal box, which totters on a brick pedestal and appears to be on the brink of collapse.

The sea wall continues beyond Dawlish and passes through five tunnels in the cliffs. This stretch of line is notorious for the lashing it takes from the winter seas and it has been breached several times. In February 1855, violent waves swept away around 100 feet of the line, leaving stranded passengers to walk with their bags across the rocks. If this sort of damage is anything to go by, then it could be that the sea wall is actually protecting the cliffs from further erosion. In the winter storms of 2013/14, the sea wall at Dawlish hit the headlines when a 130-foot

section was swept away by heavy seas on the night of 4 February, leaving the tracks floating in the air. The line was closed for repair work and reopened two months later.

There is nothing Brunellian to be found up the line at Newton Abbott station itself, which was completely rebuilt in the 1920s because of its importance as a major junction. But across the road in the compact Newton Abbott Museum there is one unexpected treasure. In pride of place in the GWR room is IKB's toilet! 'It can't be verified,' curator Felicity Cole told me back in 2005. 'It is more of a legend really. It was installed by the Earl of Devon in the engineer's office next door. Water closets were still something of a rarity when Brunel was working here, and to be honest this is a very ordinary ceramic loo with floral decoration. It's a horrible brown colour really – well let's say it is more of a ceramic beige.' I was tempted to say that the butt stopped here.

Off to the Cornish Riviera. A typical GWR publicity shot from the 1930s with Dawlish in the background.

DIDCOT RAILWAY CENTRE

LOCATION: Didcot, between Oxford, Swindon and Reading.

GETTING THERE: By train to Didcot. If driving, exit at junction 13 or 14 on the M4 and head north for Didcot. Don't be deterred by the intermittent signposts; head for station and car parks. A subway passes under the track to the Railway Centre.

OPENING TIMES: Saturday and Sunday all year (not Christmas and Boxing Day), various weekdays plus special steam days throughout the year as advertised. Times vary.

Home of the Great Western Railway Society, this large site exhibits mostly later GWR trains and buildings. Of special interest is a section of broad and mixed gauge track, the transfer shed and newly completed *Fire Fly* working replica, plus a big piece of atmospheric railway pipe. Carriage collection includes remains of a 'convertible' built for easy conversion to standard gauge, and a broad gauge carriage.

INFORMATION: 01235 817200 / www.didcotrailwaycentre.org.uk

CULHAM STATION

LOCATION: Culham, four miles north of Didcot.

GETTING THERE: Some local trains stop here. By car, it is just north of the B4016, or south from the A415 road to Abingdon.

This is a charming little station, beautifully restored and used as an office. The Railway Inn is packed with railway pictures.

FROME STATION

LOCATION: Frome, Wiltshire.
(OS map 183:784476)

GETTING THERE: ten miles south of
Bath on A36. Don't expect too much
help from street signs in the town; the
station is on the eastern side just off
the A3098, behind Jewsons builders'
merchants. You can usually park at the
station.

This pleasing wooden structure was designed by one of Brunel's assistants.
Unfortunately the fine broad gauge goods shed has gone.

STROUD STATION AND ENGINE SHED

LOCATION: Stroud, Gloucestershire.
(OS map 162: SO 850052)

GETTING THERE: Trains run from
either Swindon or Gloucester. Stroud
is situated six miles from the M4,
junction 13 from the south, or junction
12 from the north. Alternatively, take
the A46 south of Cheltenham, or the
A419 from Cirencester. Parking in
town can be difficult, although there is
a large car park near the station beside
the imposing Hill Paul building. Parking
in the station car park itself can be
expensive.

A good example of a Brunel station, the real gem is the stone-built broad gauge
goods shed. Excellent views of this from either platform – note the old GWR
advertisement still visible on its trackside wall – but don't be tempted to stray
beyond the warning signs. The best view is close up, so head out through the main
station building, turning right and along by the parking area.

ATMOSPHERIC RAILWAY, SOUTH DEVON

LOCATION: The railway line hugging the west bank of the River Exe includes Starcross and Dawlish. (OS map 192:977817 and 964767)

GETTING THERE: This is a busy railway line with some trains stopping at Dawlish, but few at Starcross. By road, take A379 south from Exeter.

LOCATION: Newton Abbot & GWR Museum, 2A St Paul's Road, Newton Abbot, TQ12 2HP.

OPENING TIMES: GWR room mid-March to October Monday–Thursday 10.00–16.00, Friday 10.00–12.00. Admission free.

You can't miss the former pumping station at Starcross, a solid brick-red Italianate building that looks like a church tower sandwiched between the road and railway. Try parking at The Atmospheric Railway pub and cross the railway via the footbridge at Starcross station. Don't get too excited by a sign directing you to 'Brunel's Atmospheric Railway' as it only indicates a good viewing point.

The pumping station at Newton Road in Torquay is in good shape – now part of the premises for the Frank H. Mann wholesalers. The one at Totnes is incorporated within a dairy.

The Newton Abbot Town & Great Western Railway Museum in St Paul's Road, Newton Abbott, has various GWR artefacts including Brunel's water closet, installed for him by the Earl of Devon.

INFORMATION: 01626 201121 / www.museum-newtonabbot.org.uk

Dawlish station is devoid of architectural merit, with the possible exception of the dilapidated signal box. There is no access to the beach side via the station itself so pass under the line beneath the low-level bridge. A short walk along the hefty sea wall buttresses leads to a footbridge on eastern side of the town for a view of the coast-hugging line complete with backcloth of deep red cliffs.

BRUNEL MANOR

LOCATION: Teignmouth Road, Torquay, Devon, TQ1 4SF.

GETTING THERE: Take the A379 coast road between Teignmouth and Torquay. Heading south, Brunel Manor is on the right-hand side after Maidencombe, four miles from Teignmouth.

This is the site of Brunel's home by the sea. He didn't design this 1850s house, but did produce drawings for the gardens. It is now a hotel specialising in Christian holidays, conferences and meetings. You can stay there, or if you just want to visit, please contact them first. IKB also designed cottages for the workers.

INFORMATION: 01803 329333 / www.brunelmanor.com

MORE BROAD GAUGE CONNECTIONS

LOCATION: South Devon Railway Trust, Buckfastleigh, Devon, TQ11 0DZ.

OPENING TIMES: Open from Easter to October.

The only surviving broad gauge engine.

The SDR runs trains between Buckfastleigh and Totnes. In the engine shed museum is *Tiny* – not designed by Brunel – a small four-wheeled vertical boiler engine used for shunting.

INFORMATION: 01364 644370 / www.southdevonrailway.co.uk

REPLICA OF THE *IRON DUKE*

LOCATION: Gloucestershire Warwickshire Railway, Toddington, Gloucestershire GL54 5DT.

OPENING TIMES: Variable – see website.

INFORMATION: 01242 621405 / www.gwsr.com

BROAD GAUGE SOCIETY

Formed in 1980 to promote the research and modelling of broad gauge railways.

INFORMATION: www.broadgauge.org.uk

Other broad gauge buildings can be found at various locations, including Ross-on-Wye and on the line from Weston-super-Mare to the south-west.

Brunel Manor stands on the site of IKB's intended retirement home.

11

TRIUMPH OVER THE TAMAR

Devon and Cornwall

Already engineer of the South Devon Railway, IKB's first involvement with the extension of the line to Falmouth and Penzance began in a purely advisory capacity. In February 1845 he appeared before a parliamentary committee examining a Bill submitted by the Cornwall Railway. Their scheme was to extend the existing coastal line from Devon with the financial backing of the so-called Associated Companies – principally the GWR and other broad gauge companies. In isolation a Cornish railway was not financially viable, but existing companies would reap the benefits of carrying additional passengers to Cornwall over their lines. It was going to be an expensive undertaking, building a line with maybe forty-three viaducts between Plymouth and Falmouth, not to mention getting over the River Tamar.

The original plan for this crossing was to load the trains on to ferries – an awkward procedure given the 18-foot rise and fall of the fast-flowing tidal waters. Curiously, IKB voiced his approval of the scheme. 'I am prepared to say that I consider there is no difficulty in doing it,' he told the committee.

Why? Well maybe he saw it as a way of ingratiating himself with the company's directors with a view to becoming their engineer when the ferry plan fell through. He must have known that a bridge was the only answer. Predictably the ferry scheme was dismissed and the directors were told to go away and reconsider their plans. This they did and, spurred on by a rival scheme to take the line further inland via Okehampton, they appointed IKB as engineer in August 1845. He immediately started a new survey of the route including a bridge at Saltash, where the river narrowed to 1,100 feet. When the revised Bill went before Parliament, the rival inland scheme had fizzled out and with Brunel's considerable expertise and reputation behind it, the Cornwall Railway Act received royal assent in August 1846. The exact form of the bridge had yet to be determined although IKB may have had a massive single span in mind, possibly of timber construction to save costs. Unfortunately

A pub sign in Saltash. Note the support shaped like part of the bridge.

Britain's first flush of railway growth suffered its first setback the following year, with a massive slump in the value of railway shares. Work on the Cornwall Railway project came to a grinding halt, leaving Brunel time to pursue his many other jobs, and a breathing space to work on ideas for the Tamar.

Chepstow Railway Bridge

By chance, IKB was faced with a similar dilemma in taking the South Wales Railway over the River Wye at Chepstow. Here the river was 300 feet wide and the Admiralty insisted on a level soffit and a clearance of at least 50 feet above the high water. To further complicate matters, the site was asymmetrical, with 120-foot-high limestone cliffs on the eastern side. You might wonder why he didn't just build a suspension bridge, but while suspension bridges cope well with the loading of nineteenth-century road traffic, they are not well suited to a substantially heavier railway train, possibly in the order of 100 tons or more. Under such weights a suspension bridge will sag, creating a ripple effect as the train passes over. So, what now? Enter IKB's old friend, Robert Stephenson.

Stephenson had similar constraints on widths and vertical clearance for a bridge to take the Chester & Holyhead Railway over the Menai Strait. In comparison with Brunel's flamboyance, Stephenson was a more cautious character, preferring incremental experimentation and research, leading to workmanlike solutions. He shunned the shock of the new for its own sake and his train of thought for Menai took him along the route of a wrought-iron box girder in the form of an iron tube big enough for trains to run inside it. The tubular spans linking three starkly functional masonry piers were straight as a die, 460 feet long and weighing 1,500 tons. Prefabricated on the shoreline, the plan was to float them on pontoons at high tide and to use the current to position them between the piers. They would then be slotted into channels running up the masonry and slowly raised into position by means of hydraulic jacks. As a mark of their mutual respect, Stephenson invited IKB to witness the installation of the first span on 20 June 1849. Knowing full well that he had only a couple of hours before the tide fell, Stephenson directed the operation from atop the girder with the smaller figure of IKB standing at his side, smoking a cigar. All went to plan and the following year Stephenson's no-frills box girder structure opened for business as the Britannia Bridge.

IKB's bridge at Chepstow also featured wrought-iron box girders, although in a tubular form, similar to those used on his little bridge in Bristol Docks. As Chepstow was on a bigger scale, he opted for a belt-and-braces approach, incorporating suspension chains within the design. On the high side of the bridge, a shallow cutting through the limestone cliff led to a pair of masonry towers pierced by arches. On the opposite side, two similar towers of cast iron stood on top of cylindrical iron piers and beyond these were three further approach spans. Between the towers were two independent bridges, in effect, side by side and each carrying a broad gauge track on longitudinal plate girders. Suspension chains curved down from the towers, while overhead two slightly arched wrought-iron tubular girders, 9 feet in diameter, counteracted the compression forces. In engineering terms, this is a closed system with all the forces of compression and tension contained within the structure. To provide rigidity, further vertical struts and diagonal ties linked the tubular girders with the deck beams. The first line of track was operational by July 1852 and the other was completed the following year. According to John Binding, an expert on Brunel's tubular girder bridges, the cost of the Chepstow Bridge, with a nominal span of 600 feet, was '£77,000 compared with £145,190 for Stephenson's 400-foot Conway Bridge'. Sadly the railway bridge at Chepstow made way for a newer bridge in 1962 and all that remains today are the stumps of the original piers.

The railway at Chepstow gave Brunel the opportunity to test his cylindrical box girders on a larger scale. It was an awkward, asymmetrical site with cliffs on one side.

The Bridge at Saltash

Returning to the Saltash project, IKB settled upon a design featuring two spans of 465 feet, reduced to 455 feet later, with a single mid-river pier. On dry land on either side would be piers linked by a series of shorter spans. In order to investigate the nature of the riverbed for that central pier, in 1848 he had a wrought-iron tube constructed, which was 85 feet long and 6 feet in diameter. Acting like a cofferdam, this was lowered with one end resting on the riverbed so that a series of trial borings could be made. Fortunately this revealed a bed of very hard rock beneath the thick layer of mud and slime. Armed with this information and his experiences at Menai and Chepstow, he was able to formulate his design for Saltash, but yet again a lack of funding put matters on hold for a further three years. In 1851, Brunel attempted to resuscitate the Cornwall Railway project, proposing that the line from Plymouth to Falmouth, including the Saltash bridge, should be reduced to a single track, thus making considerable savings. This was approved in 1852, by which time he had finalised his design for the bridge and the result is, to say the least, spectacular.

Like any great journey, my travels through the life's work of IKB have been full of surprises and unexpected moments of wonder. Perhaps the greatest of these occurred on the approach to the Tamar estuary at Saltash with the first glimpse of the Royal Albert Bridge. As you drive from the Plymouth side, the grey humps of tubular iron rear up from the river valley like the back of a vast serpent. The 1961 road bridge running alongside might dwarf Brunel's rail bridge, but it does not diminish the old bridge's commanding presence – a magnificent marriage of engineering and landscape.

As with any bridge, the visitor is spoilt for choice when it comes to choosing where to start. On the Devon side there is a car park area that overlooks the bridge with a view straight down its single track. But I chose to bypass this and head straight for Saltash – perhaps drawn by the name but also by the presence of the old railway station. The town itself is unremarkable, although I had good reason to appreciate the shelter of the Brunel pub when the slate-grey sky let rip with a sudden torrent of hailstones. Before getting up close and personal with the bridge, I wanted to head down to the river's edge and take in the old station on the way. The squat station building was boarded up and in a sad state of neglect and there were plans to convert it into a visitor centre at some point. From the platform there is an excellent view along the curving track to the bridge, one that is familiar from many of the old engravings. The station sign features a colourful image of the bridge above the slogan 'Welcome to Cornwall', or '*Kernow a'gas dynnergh*' if you prefer the local parlance.

The walk to the water's edge is steep, but once there you really get an impression of the scale of this bridge. In old photographs it is shown looming above the narrow streets, but many of the buildings are now gone, leaving a scattering of pubs, a children's playground in the form of a broad gauge train and the boatyards, where small pleasure boats sit out the tides. Looking up to the railway bridge, you can't fail to notice the Tamar Road Bridge on the other side. This modern suspension bridge has drastically reduced the travel time into Cornwall and whatever your opinion regarding its impact on its neighbour, it does at least allow for a closer look at Brunel's masterpiece. There is a toll for vehicles, but pedestrians don't pay. Incidentally, it is said that in the past many a resident of Saltash took the dangerous walk across IKB's rail bridge having missed the ferry after a night in Plymouth, and during the Second World War wooden slats were installed between the rails in case military vehicles needed to make a quick crossing.

Having climbed the hill to the Tamar Road Bridge, I was nearly swept off my feet by the howling gale blowing down the estuary – the whole structure seemed to be swaying. It is a surprisingly long walk from one

The dilapidated station building at Saltash may be converted into a visitor's centre in the near future.

side to the other, especially if you intend clambering up and down to the riverbank, but my first stop was the midway point to look at Brunel's central pier and to start making sense of his design. This bridge has been described as a 'masterpiece of complexity' and looking at those great hump-backed spans, it is easy to see why. As with Chepstow's bridge, it is a closed structure incorporating three engineering forms; the compression arch, the tension chains of a suspension bridge, and a beam deck. If you blot out the tubular arches, it is easier to make out the curve of the suspension chains between the towers just as on the Clifton Bridge. In fact, some of these chains came from Bristol when that bridge was unfinished, and the shape of the towers bears more than a resemblance to their Clifton counterparts. However, the difference at Clifton is that the chains pass over the towers and are anchored into rock on either side creating an 'open' structure where the forces are equalised. With the double span at Saltash, there is nowhere for the chains to go and so their tension forces, pulling the towers inwards, are countered by the arched tubular girders on the top. This is described as a 'bowstring' design. To stiffen the structure, IKB incorporated vertical struts running from the girders to the deck with further diagonal bracing for longitudinal stability.

The first task in building the bridge was to erect the central pier. The main part above the waterline consists of four octagonal cast-iron columns 96 feet high, but beneath them is a masonry column going a further 96 feet down into the water. To construct this, IKB built another steel cylinder, this time 35 feet in diameter at the base, creating what in effect was a diving bell in which the stonemasons could work. To keep the river from seeping in, the air inside was kept under pressure by steam pumps and this made it possible for men to work 70 feet under the water, although they frequently suffered severe headaches and muscular pains as a result of what we now term 'the bends'. The central pier was complete by the end of 1856 and the cylinder was unbolted, split into two and removed. Construction of the land piers for the two main spans and for the seventeen approach spans – eight for Devon and nine for Cornwall – was a far simpler matter. The portal towers at either end feature a masonry and brick lining within a cast-iron shell, while the central one is of cast iron.

Meanwhile, the first of the two trusses was taking shape on the Devon shore. The upper main girders, unlike the circular tubular girders at Chepstow, were oval in cross section, being 12 feet 3 inches high and 16 feet 9 inches across. This served the dual purpose of reducing sideways

During construction, the scene at Saltash from the Devon shore.

wind resistance and also permitting the deck struts to hang vertically (whereas they were inclined at Chepstow). Each truss weighing 1,000 tons was to be positioned between the piers by floating it on huge pontoons, which could be raised or lowered by pumping water in or out of them. The day for moving the first truss to the Cornwall side was set for 1 September 1857. IKB made meticulous plans and took command from a platform mounted high up on the truss accompanied by a band of flag-waving signallers to convey his orders. Thousands of onlookers, estimates suggest 300,000, crammed the shoreline and surrounding hills to witness this wonder, but they had been warned that absolute silence was demanded by the great engineer. As the signal flags fluttered out their instructions, an eyewitness recorded the event:

Not a voice was heard, not a direction spoken; a few flags waved, a few boards with numbers on them were exhibited, and, by some mysterious agency, the tube and rail borne on pontoons travelled to their resting-place, and with such quietude as marked the building of Soloman's temple. With the impressive silence which is the highest evidence of power, it slid, as it were, into position without an accident, without any extraordinary effort, without a 'misfit' to the extent of the eighth of an inch. Equally without haste and without delay, just as the tide reached its limit, at 3 o'clock, the tube was fixed on the piers 30 feet above high water …

This was Brunel's moment of triumph. Here was the master magician in complete control, the 'Little Giant' reshaping the landscape with his massive iron construction. When the band struck up 'See the conquering hero comes', the spell of silence was broken and the vast crowd suddenly erupted into a cacophony of cheering.

The process of raising the truss was a much more protracted matter, with huge hydraulic jacks lifting it 3 feet at a time as cast-iron sections were added to the central pier and the masonry was constructed on the land pier. This had to be given time to set between each lift and so it wasn't until May the following year that it was in its final position. By the time the second span was ready to be moved, Brunel was in London struggling with the *Great Eastern*. His chief assistant Robert Brereton supervised the flotation and positioning in July 1858, and it was under his guidance that the bridge was finished in early 1859. When Prince Albert travelled down from Paddington to open the Royal Albert Bridge in May of 1859, IKB was too ill to attend. Shortly afterwards he saw the completed bridge for the first and the last time. Reclining on a couch placed on an open trolley, he was drawn slowly across by one of Daniel Gooch's steam locomotives. 'No flags flew, no bands played, no crowds cheered …'

Brunel died a few months later in September 1859. As a tribute, the Cornwall Railway directors added a simple epitaph to the portals at each end of the bridge. 'I.K. BRUNEL ENGINEER 1859'. Today the Royal Albert Bridge at Saltash is little changed and is in continual use, although even the high-speed HST 125s are humbled by the 15mph speed limit. The track was removed and there have been a few minor modifications to the structure, including the addition of horizontal girders to strengthen the main spans, and the wrought iron of the approaches have been replaced by steel. In 2014, a major £10 million refurbishment of the bridge was completed by Network Rail.

This tale of the great railway bridges has been one of rivalry and friendship between the two men of iron, Brunel and Stephenson. In stark contrast to his friend's cautious approach, IKB rode the leading edge of engineering, embracing the challenge of the new and revelling in the inherent risks in pursuit of his own driving ambition to be the greatest engineer that ever lived. There is only one Saltash-type bridge. But then, there was only one engineer the likes of Isambard Kingdom Brunel.

Cornwall's Wooden Viaducts

Beyond the Tamar, the course of the Cornwall Railway crosses many deep valleys. Because of the economic constraints and the lack of raw materials from cuttings to create embankments, IKB constructed a series of more or less standardised viaducts either wholly of timber or a combination of timber and stone. And if the Royal Albert Bridge impressed you, then what remains of Brunel's lofty Cornish viaducts will not disappoint.

Work on these viaducts had been carried out at the same time as the Saltash bridge so when that opened, the route from Paddington all the way to Truro was complete. Except where passing over wet or marshy ground, the viaducts consisted of masonry piers 66 feet apart and rising up to within 35 feet of the rail level. The piers were capped with iron plates from which four sets of heavy timbers radiated like an upturned hand, with further cross-bracing, supporting longitudinal beams beneath the deck. It is said that these viaducts fitted in perfectly with the 'primeval, storm-bitten landscape of western Cornwall'. Maybe so, but many of the locals were afraid to travel across the flimsy-looking structures. Admittedly it must have taken some nerve to cross a wooden viaduct possibly 153 feet high – such as the one at St Pinnock – on a dark windy night.

By the time the line was doubled in 1908, the viaducts on the main line to Penzance had been replaced either with masonry extensions and steel girders sitting on the old piers or with complete new masonry viaducts, although some on branch lines survived a little longer. IKB had never intended them to last more than thirty years (whereas some lasted for sixty) and had designed them with standardised lengths of timber for easy refurbishment. What killed them off was the need to widen the line combined with a sharp rise in timber prices, in particular for Baltic pine.

It is impossible to cover all the viaducts in detail here, but a good selection can be found within easy reach of Saltash. The first viaduct is just eight miles along the A38 at Lower Clicker, situated to the north of the road which is probably the best place from which to view it. The sign indicating Menheniot station took me through the little village, where I discovered the Sportsman's Arms Hotel right next to the western end of the viaduct. This is only a stone's throw from the track, but from so close up you literally can't see the viaduct for the trees. It is possible to stay with the road, passing under the railway and down into the valley in order to loop back towards the viaduct. Here, I encountered a gate festooned with signs pointing out that the rest of the muddy track is a private road, and while not exactly suggesting that all trespassers will be shot on sight, it did indicate that you could be crushed by runaway tree trunks stacked by the foresters. Even at a distance this is an impressive viaduct; Brunel's timberwork is gone of course, replaced by a masonry extension and capped with a steel girder bridge. When viewed from the muddy path underneath, the slender stone piers – buttressed and pierced with Gothic arches – seem to rise up forever.

My next destination was a few miles to the west: the ancient market town of Liskeard and its station and viaduct. The road winds its way right around the town before you reach the station on the south side. At first sight, the cream-painted building with its wooden canopy promises something very traditional. However, this is a working station and it has several modern additions, including a glass-fronted cafe. Don't be put off, because it is still a charming station with its wide 'flying bridge' springing direct from the slopes of the cutting to carry road traffic. On the platform new shelters mingle with the old, and note the marvellous old wooden signal box which is still manned, though not of Brunel

vintage. (Around the corner is another small station building and a single-track line at right angles to the mainline. This runs to the seaside resort at Looe a few miles to the south.) Liskeard's viaduct, immediately south-east of the station, is one of the deepest in Cornwall and, at 720 feet long, one of the longest. It features masonry extensions to the original piers and girders.

From Liskeard I headed west and found the Moorswater viaduct. This is another big one and you can't miss it from the road. These structures were never built as visitor attractions, and it can be hard to get a good overall view of the tall stone piers, which stand in the shadow of the newer masonry viaduct like a row of ancient castle towers. Arched piercings add to this impression and as the piers slowly crumble, they are being reclaimed by the landscape.

It should be noted that Brunel used timber on viaducts on a number of other lines, most notably including the Bristol & Exeter Railway, Bristol & Gloucester Railway, Cheltenham & Great Western Railway, South Devon Railway, South Wales Railway, on the Oxford to Birmingham lines, and on several other others. These wooden viaducts have all gone now, but there are several good books on the subject. See *Brunel's Cornish Viaducts* by John Binding, *Brunel's Timber Bridges and Viaducts* by Brian Lewis, and *Brunel in Cornwall* by John Christopher. Details are given in the bibliography.

Photograph of the timber-built Carvedras viaduct at Truro. Later replaced by a masonry viaduct in 1905, five of the original masonry piers have survived.

ROYAL ALBERT BRIDGE, SALTASH

LOCATION: The bridge spans the Tamar between Plymouth on the Devon side and Saltash in Cornwall. (OS map 201:436587)

GETTING THERE: Trains to Saltash are infrequent, so Plymouth is the nearest. By road you are spoilt for choice with parking on the Devon side overlooking the bridge, or head into Saltash and the town centre.

This is arguably Brunel's greatest bridge, so take your time. You get a wonderful impression of its scale from river level with a scattering of pubs on the Saltash side or visit The Brunel up the hill. Saltash station is between the town centre and riverside and in a sad state when I visited, but there were plans to make this a new Brunel Heritage Centre. The 1961 Tamar Road Bridge provides an excellent close-up view of Brunel's bridge. Pedestrians travel free. You tend to be shooting into the sun during the middle part of the day, so try putting your camera away and imagine the ailing Brunel viewing his newly completed masterpiece from an open-topped carriage.

INFORMATION: www.saltash.gov.uk or www.royalalbertbridge.co.uk

The Royal Albert Bridge seen from the platform at Saltash station. Note how the track reduces to a single line.

12

LEVIATHAN

Millwall, London: The *Great Eastern*

One of IKB's greatest talents was in what we now term multi-tasking, and while he was dealing with the extensions to the broad gauge network, he was also working on the 'Great Iron Ship'. It was to have been the culmination of his brilliant career, but instead it killed him.

Given the financial track record of his previous steamship, which through no fault of his had bankrupted the Great Western Steamship Co., you might have expected Brunel to have abandoned his nautical ambitions. In 1851 he advised the directors of the Australian Royal Mail Co. that a ship of between 5,000 and 6,000 tons was just what they needed for the Australian mail contract. They rejected this suggestion and instead invited him to tender for a pair of ships, approximately the size of the *Great Britain*, which would have taken on coal en route at the Cape of Good Hope. These were never built and IKB's thoughts turned to bigger things. In March 1852 he sketched a rough design for a massive steamship under the heading 'East India Steamship'. In the accompanying notes he reveals an almost casual approach in tackling the largest movable object the world had ever seen: 'Say 600 feet by 65 feet by 30 feet.' When completed, it was 692 feet long with a gross tonnage of 18,915 tons; almost six times bigger than anything afloat.

The attraction in creating such a vessel was the burgeoning and lucrative Australia route (which the *Great Britain* was servicing from 1852 as a sailing boat). What was needed, he concluded, was a steamship big enough to travel to Australia and back without refuelling. This was based largely on the premise – a false one as it happens – that there was no coal in Australia. But his vision was also flawed on the practical and commercial levels. The world wasn't ready for such a big ship. It wouldn't fit into most harbours and accordingly, at one stage, IKB envisioned a pair of feeder boats to shuttle passengers to the shore. Because emigrants are by nature a one-way cargo, it was suggested that goods could be brought back from India on the return leg; even though a ship of this capacity would carry so much cargo it would probably swamp the market and actually serve to depress prices. Nevertheless, in typical Brunellian style, he managed to sell the concept to the directors of the Eastern Steam Navigation Co., but it was not to be plain sailing. In March of the following year the government awarded the mail contract to the Peninsular & Oriental Steam Navigation Co. There were resignations among the directors and investors became thin on the ground; even Brunel's charm offensive failed to persuade his many friends and contacts to put up enough money. Uncharacteristically, he decided to take a large financial stake in the company himself:

I then set to work to get the capital and by December 1853 after great exertion 40,000 shares were taken and £120,000 paid up … More than once we have nearly failed and broken down … After two years' exertions we are thus set going, contracts entered into and work commenced 25 February 1854.

Construction

The shipbuilder John Scott Russell had submitted the winning tender to construct the as yet unnamed ship. IKB had estimated its cost at £500,000 but Scott Russell's figure was £377,200 and he proposed reducing this further if he received a contract for an identical sister ship. IKB wasn't exactly known for underestimating costs and it seems extraordinary that Scott Russell was so keen to get the job that he was offering such a low price without any scope for unforeseen contingencies. Regardless of the money, he couldn't have anticipated the difficulties that lay ahead working with, or perhaps under, the little engineer with the big ego. It was a recipe for disaster.

Over the next five long years, thousands of sightseers flocked to Scott Russell's shipyard at Millwall on London's Isle of Dogs. What they saw, rising like a cliff of iron above the marshy fields and ramshackle buildings, was a vessel described by many as the first modern ship. And while Scott Russell may have contributed much to the design, the overall concept was undoubtedly pure Brunel. 'The whole of the vessel is divided transversely, into ten perfectly water-tight compartments, by bulkheads carried up to the upper deck. So that if the ship were to be cut into two, the separate portions would float.'

In addition to the watertight compartments, the hull was constructed with a cellular double skin with iron plates 2 feet 10 inches apart and containing longitudinal members spaced at 6-foot intervals. When combined with the upper deck, also of cellular construction, and two 36-foot-deep longitudinal bulkheads, it created what in effect was one huge box girder, making the ship immensely strong. Instead of treating the iron construction as a natural progression from wooden designs, IKB had returned to first principles with the result that every piece of ironwork served to strengthen the hull structure:

> In the present construction of iron ships … nearly twenty per
> cent of the total weight is expended in angle irons or frames
> which may be useful or convenient in the mere putting together
> of the whole as a great box, but is almost useless, or very much
> misapplied, in affecting the strength of the structure of the ship. All
> this misconstruction I forbid, and the consequence is that every
> part has to be considered and designed as if an iron ship had never
> before been built; indeed I believe we should get on much quicker
> if we had no previous habits and prejudices on the subject.

In the light of his experience with the *Great Britain* and a series of tests he had conducted on behalf of the Admiralty, IKB wanted a propeller to drive the new ship. But this presented overwhelming practical problems with such a huge vessel; in particular with building an engine that would be powerful enough and with forging the propeller shaft. There were also limitations with the size of the screw, as a shallow draught would cause a large propeller to rise out of the water. Paddle wheels might be less efficient but at least they permitted the ship to navigate shallower waters, and the combination of the two systems promised an expected speed of 15 knots and improved manoeuvrability. Of course it also meant that two engines and boiler rooms would be needed. The 24-foot diameter four-bladed screw was to provide approximately 60 per cent of the propulsion and was driven by a 1,600hp four-cylinder horizontal direct-acting steam engine. For the 56-foot diameter paddle wheels, a 1,000hp four-cylinder oscillating engine was sufficient.

If you go down to the Millwall area beside the Thames, there is just enough to evoke something of the titanic struggle that played out here. The site isn't that easy to find and the only clue on my street map of London was the Great Eastern Pier, shown on the south-west corner of the Isle of Dogs. The area has been redeveloped in recent years, mostly for housing, but situated at Burrell's Wharf Square, a section of the ship's slipway has been uncovered. The standard way to launch a ship is stern first, but this ship was far too long for the relatively narrow Thames and it would probably have rammed the opposite shore. Originally IKB had advocated a dry dock for her construction, as with the *Great Britain*, but this was deemed too expensive given the gravelly ground conditions. So instead he decided to flout convention and build the ship sideways to the river and then slide it into the water on cradles riding on wooden launch ways. This method wasn't entirely without precedence – it had been employed by a shipbuilder in the USA with some success, but never on such a vast scale. To support the ship and the launch ways, 1,500 oak piles were driven 24 feet into the ground, and across them a grid of supporting timbers sloped gently down to the water – in all, 600 tons of timber were used. Each of the hefty timbers is about one foot square in cross section and, as I clambered about them, it was not difficult to imagine IKB posing for Robert Howlett's famous photographs, mud splattered on his trousers and a backdrop of gigantic chain links.

In appearance, the *Great Eastern* had severe lines with no superstructure above the main deck and no upward curve at the front or back as with

One of the Tangye hydraulic rams used to push the ship into the Thames.

His unremitting workload, years of chain-smoking cigars and too little sleep were finally taking their toll. He was also suffering from Bright's disease of the kidneys, and consequently his judgment was often questionable and his frame of mind inclined toward paranoia. His exchanges with Scott Russell became increasingly irascible and he ended one letter to him with, 'I wish you were my obedient servant, I should begin by a little flogging …'

Launching the *Great Eastern*

Even for the great IKB, this massive project was beyond his mastery, and for once in his life circumstances were dominating him. But he couldn't stop, and once he had completed the ship, he still had to shift it. Opposing the idea of a free launch, he insisted that the ship's movement had to be under his full control. Large wooden cradles were constructed, with their lower edges faced with iron strips, and in order to regulate her descent, each cradle was connected via chains to huge checking drums on the slipway.

The launch was due to take place on 3 November 1857, a date selected according to the tide. IKB had demanded absolute silence for the proceedings, and to convey his instructions he adopted the same system of signals deployed at Saltash only a couple of months earlier. But to his absolute horror, he found the shipyard swarming with thousands of rowdy onlookers. The company directors had sold 3,000 tickets to the spectacle in a bid to recoup some of their money – a portent of things to come for the ship. One eyewitness recorded the scene as a pall of grey drizzle fell upon them:

most ships. The hull was constructed from standardised iron plates and angle irons – unusual in itself at the time – and as there were no cranes, each plate had to be hauled into position and riveted. L.T.C. Rolt described the scene: 'Every one of millions of rivets had to be closed by hand, the rivet boys and holders-up labouring for months on end amid a deafening clangour in the confined space between the hulls.'

As work proceeded, behind the scenes the rivalry between engineer and shipbuilder intensified into a sorry tale of two prima donnas vying for the limelight. Traditionally Brunel's biographers have cast Scott Russell as the villain in the ensuing drama, going so far as to describe him as an 'evil genius'. But in truth, they were as bad as each other. On the one hand, IKB would constantly modify his designs, while on the other he was looking over the shoulder of Russell and his workforce, and would complain vociferously when they failed to keep pace with his expectations. He also held the purse strings and only paid Scott Russell as the work proceeded.

It didn't help that their manoeuvrings were conducted in the full glare of publicity. Brunel became incensed by newspaper reports that sidelined his contribution to the ship's design and, increasingly irascible owing to painful bouts of illness and failing health, he struck out against his adversary with venom. In February 1856, matters came to a head with the work behind schedule and, according to Brunel, a substantial quantity of iron plate unaccounted for. It was all becoming too much for him: 'I never embarked on any one thing to which I have so entirely devoted myself, and to which I have devoted so much time, thought and labour, on the success of which I have staked so much reputation.'

> Across the narrow streets, from public-house to public-house, were stretched broad flowing flags, and every apartment in every house, whether bedroom or sitting-room, if it commanded even a glimpse of the huge vessel stretching along above the tree tops, was turned 'inside out' to accommodate visitors for friendship, relationships or lucre. Bands of musicians were enlivening the scene at the different public-houses, even at the early hour of ten in the morning, and the performers were drunk at that period, and miserably out of time and tune …

Until now the ship had been known simply as the 'Great Ship' and the company directors produced a list of names for the launching ceremony.

'You can call it Tom Thumb for all I care,' retorted Brunel. But instead they chose *Leviathan*, a name that did not stick; she soon became known as the *Great Eastern*.

When the time arrived for the steam winches to start, there arose a 'rumbling noise like a prolonged roll of drums' as the straining chains reverberated against the empty iron hull. The ship juddered. The bow jerked forward, taking up slack on the rear checking drum and causing the long wooden handle of the drum to spin around, tossing workers aside and killing one labourer. In the ensuing pandemonium, Brunel halted the launch. The cheers turned to jeers and the newspapers openly taunted him. His moment of triumph had become a humiliating and public failure. 'Could I have foreseen the work I have to go through, I would never have entered upon it, but I never flinch, and do it we will.'

On the next attempt, on 19 November 1858, the ship moved only 14 feet. Exasperated, Brunel once again turned to Robert Stephenson for help. Stephenson, also in poor health, could not refuse his close friend in his hour of need and he left his sickbed to make his way to Millwall. Together they realised that only brute force could move the monster and the country was scoured for hydraulics jacks that could push the ship, inch by inch, into the Thames. Moving at a snail's pace, it took three months before she was in the water and on 31 January 1859 the ship floated above the wooden cradles.

After five years of work, Brunel was utterly exhausted and he was seriously ill, suffering from Bright's disease – a kidney disease better known as chronic nephritis today. On 5 September, two days before her maiden voyage, he took one last look around the ship. On the deck he posed for a photograph beside one of the tall funnels, a walking stick in one hand and his stovepipe hat in the other. The image is of a frail and broken man aged before his years. Moments afterwards, Brunel slumped to the deck – some sources say with a stroke and others a heart attack – and he was carried off the ship on a stretcher.

The ship sailed without him, and a few days later he received some terrible news. As the *Great Eastern* sailed down the English Channel, valves that should have been left open had been closed, causing a feed-water heater at the base of the forward funnel to explode, sending it shooting 30 feet into the air like a rocket. This caused the forward boiler to blow back and suddenly through clouds of steam and smoke, several hideously scalded stokers rushed on to the deck, some leaping over the side. The accident claimed five lives and for the ailing Brunel this was unbearable news. He died at his Duke Street home on the

The modern lines of the *Great Eastern* as she rounded the point opposite Blackwall in September 1859. Note the huge sea anchors.

A marvel of engineering, the paddle-engine room of the *Great Eastern*, depicted by the *Illustrated London News* in September 1859.

evening of 15 September. Stephenson passed away the following month. The 'heroic age of engineering' was over and like some operatic tragedy, the men of iron had been consumed by their own creations.

The day after the explosion the *Great Eastern* docked at Weymouth for repairs to the funnel and to minor damage in the forward saloon.

In June the following year she made her first trip to New York, but the ship made only a handful of voyages as a passenger ship. Instead she became little more than a tramp steamer and a curiosity for the paying public. Later on she began a successful career as a cable-layer when she was chartered to the Telegraph Construction Co. Her passenger accommodation was removed and the vast hull converted to hold thousands of miles of cable. Described as 'an elephant spinning a cobweb', she laid the first transatlantic cable between Britain and the USA in July 1866, and she went on to lay cables from France to America, Bombay to Aden and up the Red Sea. When she was replaced with a custom-built cable-laying ship, *Great Eastern* ended her days as a

Brunel posed beside one of the ship's funnels for this final photograph, taken moments before he collapsed and was carried off the ship. A far cry from the famous chains photograph taken less than a year earlier.

The moment when the cable broke on the *Great Eastern*'s first attempt to lay a transatlantic cable, July 1865.

showboat anchored in the Mersey. In November 1888 she was sold for scrap. But it is said that the old girl did not give the breakers an easy time, so well was she built.

Precious little remains of the *Great Eastern*, although one relic has recently come to light. When she had put in to Weymouth for repairs, a section of the old funnel was acquired by the water company and for over 140 years it served as a filter for the town's water supply. In 2003 the funnel was donated by Wessex Water to the SS Great Britain Trust, where it is displayed a few yards away from the *Great Eastern*'s steam whistle.

London: Kensal Green Cemetery

IKB was buried at the Kensal Green Cemetery on 20 September 1859. Some visitors to the grave are disappointed by the modest stone or question this sprawling cemetery as a fitting resting place for the great man. They forget that this was a very fashionable cemetery in Victorian times and this is a family grave where his parents were already buried. It was once a place where you would have heard the trains passing by on the Paddington to Bristol mainline just to the south.

The *Great Eastern* ended her days beached in the Mersey. (Campbell McCutcheon)

LAUNCH SITE OF THE *GREAT EASTERN*

LOCATION: Millwall, Isle of Dogs.

GETTING THERE: Alight at Mudchute or Island Gardens station on the Dockland Light Railway. By car, take the Westferry Road down western side of the Isle of Dogs and turn right into Napier Avenue. Residents-only parking, but there is usually space.

Huge timber supports remain from the slipway for the ship, and at low tide you can see more timbers on the shore. (Relics from the *Great Eastern* can also be seen at the SS *Great Britain* in Bristol, or models and paintings at various museums.)

FAMILY GRAVE

LOCATION: Kensal Green Cemetery, W10 4RA.

GETTING THERE: Follow the A404 Harrow Road, or take the Bakerloo Line to Kensal Green station. Parking is difficult. Stay in Sainsbury's car park for up to two hours and walk to the cemetery over the canal bridge.

Sir Marc Brunel had a hand in designing this large cemetery, so it is only fitting that the family grave is here. The modest headstone is halfway between the eastern main entrance and the chapel – take the main pathway until it splits into three. It is on your left-hand side, behind a tree a few yards from path.

INFORMATION: www.kensalgreencemetery.com

13

BRUNEL'S LEGACY

In Brunel's Footsteps

Returning to London completes the circle. This is where it had all begun under the Thames and this is where it ended at Millwall. So, what had I gained from my journey through Brunel's life and works? Certainly an appreciation of his immense energy and drive, for one thing. My travels had taken me to many places I might otherwise have never sought out. My quest gave me an insight into the scale of IKB's endeavours and of the rich industrial heritage of this country. So many examples of his work have survived, but let's not overlook those that have been lost such as Hungerford Bridge, the wooden railway bridge at Bath, all of Cornwall's lofty viaducts, the broad gauge and atmospheric railways, plus two out of three of his great ships. Incredibly, some have vanished in relatively recent times, including the Chepstow Railway Bridge and the engine shed at Frome, and others may still be under threat.

Had I encountered the ghost of IKB – that ethereal figure draped in swathes of cloth walking along the uprooted broad gauge tracks? To be honest, no. Not at his grave at least. Or at his high-profile survivors. If I did come near to making contact of any sort it was at Millwall on a cold drizzly day, standing on those slipway timbers and finding a length of heavy chain strewn across the pavement. If any location can be haunted by the emotions and turmoil of a former time, then it is this place, which witnessed Brunel's last heroic struggle with the Great Iron Ship.

Forever 'hung in chains'

Another of IKB's enduring legacies is his image. As one historian put it, 'The ultimate icon of Victorian Britain turned out to be not a portrait of the great white queen but a grainy photograph of Isambard Kingdom Brunel.' That single image of him posed against the chains has aroused as much controversy and interest as any of his engineering works. In all there are nine known photographic images of IKB, of which two are studio shots, one was taken on the ship days before his death, and six date from the time of the attempted launches of the *Great Eastern*. Three of the latter are informal groups in the style of modern news photographs (possibly by the photographer George Downs) while the remaining three portraits shots are by Robert Howlett working on behalf of *The Illustrated Times*. For these he posed IKB in front of the chains of one of the checking drums. In two photographs he is standing and in the other he is seated and, judging by his clothing, this one was taken on a separate occasion.

Much has been written about the most famous 'chain' photograph. Some commentators suggest that IKB was a master of self-image and consciously set up this scene to portray the confident engineer, the little man in control of the giant chains. They often quote Brunel's comment: 'I alone am hung in chains.' But paradoxically, it takes the full quotation to tell the real story: 'I asked Mr Lenox to stand with me, but he would

Temple Meads, reflected in the face of Brunel at the Reckless Engineer pub in Bristol.

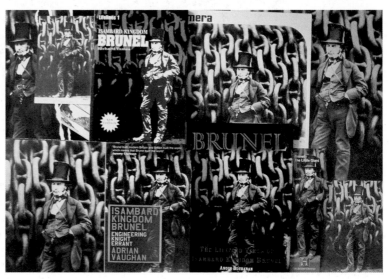

One of the most recognisable images of the nineteenth century; the photograph of IKB against the chains is everywhere.

not, so I alone am hung in chains.' So it seems that this iconic image was the result of a happy accident. Howlett had only a few minutes to capture the moment, and if IKB had been joined by a colleague, then the image would lose its impact (indeed, the other – less widely published – chain photograph is far weaker, showing him lounging like 'a yobbo waiting for a bus'). This extraordinary portrait is not of a man obsessed with his own image; this is someone whose thoughts are far away, grappling with the demons of failure and humiliation. That it speaks volumes is self-evident in its widespread recognition and its enduring power to sell the legend of Brunel, just as it sells countless magazine, postcards, books and souvenirs.

Mention must be made of that cigar. IKB is said to have smoked forty a day, which he carried in a special pouch; incidentally not on the strap shown in the photograph! The 'yobbo' photo reveals this to be his binoculars. There was an outcry when a school poster featured the photograph with the cigar airbrushed out in the name of political correctness. But in 2006 the cigar made a comeback and appeared in a comic book version of Brunel's life issued to school children to mark the 200th anniversary, accompanied by explanatory notes on the evils of smoking!

In addition to his image, Brunel's name has been appropriated by countless companies and organisations. You can drink at The Brunel wine bar in Clifton, or at the Reckless Engineer opposite Temple Meads station – perhaps enjoying a glass of IKB beer – wear Brunel shoes, go shopping in the Brunel Centre, send your children to Brunel University, warm yourself beside a Brunel wood-burning stove … and the list goes on. And you might live in Brunel Way, Brunel Close, Brunel Walk or Brunel Place in London, Bristol or a dozen other towns.

A Great Briton

It is said that the sign of greatness is the ability to inspire greatness in others. Brunel's most important legacy is the inspiration he gives to future generations through his vision and his indefatigable sense of self-belief. It is something he touched upon when he wrote these comments at the start of work upon the *Great Eastern*:

> The wisest and safest plan in striking out a new path is to go straight in the direction we believe to be right, disregarding the small impedimenta which may appear to be in our way – to design everything in the first instance for the best possible results

… and without yielding in the least to any prejudices now existing … or any fear of the consequences.

Daniel Gooch expressed it another way in his tribute to IKB:

> By his death the greatest of England's engineers was lost, the man with the greatest originality of thought and power of execution, bold in his plans but right. The commercial world thought him extravagant; but although he was so, great things are not done by those who sit down and count the cost of every thought and act.

The life, work and spirit of Isambard Kingdom Brunel continue to inspire, and when Concorde made its final flight to Bristol in 2003, it made a special flypast over the Clifton Suspension Bridge – a tribute from a modern masterpiece of high-tech engineering to the man who shaped our world.

Far left: A tribute from a modern marvel of engineering to the man who shaped our world. Concorde made this special flypast on its final flight returning to its Filton home in Bristol. (South West News Agency)

Left: A bust of Brunel, photographed in storage in 2006.

ROUND TRIPS

Many of the sites covered in this book can be grouped together to form round trips, which should be manageable in a single day, or combine these for longer excursions. Don't forget that some locations, or the land providing access, are privately owned and not necessarily open to the public. Access to railway sites is often restricted for safety reasons.

LONDON

Central London

The statue at Temple Place leads via the Embankment to the Hungerford and Golden Jubilee Bridges; walk over these to South Bank and the London Eye. From there it is not far to Westminster Abbey, either back over the footbridges or across Westminster Bridge. An additional loop takes in the National Portrait Gallery.

East London

The *Great Eastern* slipway and Thames Tunnel, along with the Brunel Engine House, are in East London. Starting at Rotherhithe, take the East London Line to Canada Water, then Jubilee Line to Canary Wharf and south on the Docklands Light Railway to Mudchute or Isand Gardens for the *Great Eastern* site at Napier Avenue. Two more stops on the DLR, or a walk through the foot tunnel, to Cutty Sark station and National Maritime Museum.

West London

Starting at the Science Museum, take the District Line to Hammersmith then north on Hammersmith & City to Paddington. Access to the Underground is by the modern glass-walled 'The Lawn', and it is three hops on the Bakerloo Line to Kensal Green cemetery – unless you decide to take a train and head off west!

OXFORDSHIRE AND BERKSHIRE

Didcot, Steventon, Culham, Mouslford and Basildon

A fairly tight group of sites can be visited with Didcot at the centre, or alternatively start from one end. East to west for example: the brick viaduct over the Thames at Basildon is visible from the A329 one mile south of Goring, but closer access is difficult. Two miles north of Goring is the Moulsford viaduct. From here the A329 leads to the A4130 and Didcot, but pause at the Moulsford railway bridge and look south-east to see Brunel's Hotel for Wallingford Road station on the north side. After the Didcot Railway Centre, Culham station is about four miles north via the B4016. Then from Culham go to Steventon, five miles or so on western side of A34.

Sonning and Maidenhead

Sonning Cutting and the Maidenhead Bridge are approximately nine miles apart on the A4 between Reading and Slough. The bridge at Windsor is between them, to the north of the town.

For a really full day, combine these two round trips into one. Basildon is twelve miles west of Sonning, and Didcot is twenty miles away.

SWINDON

Steam museum, Railway Village and Brunel statue

Come out of the Steam museum, turn right and head eastwards under the railway line through a long pedestrian tunnel. This comes out on to the Village. Note how close some people had to live to their workplace. Continue across the busy main road and into the Brunel Shopping Centre area for the statue.

BRISTOL

These two tours are easily combined – begin at Clifton or the city centre – a longish walk between the SS *Great Britain* and the docks, and a steep climb up to Clifton.

Central Bristol

This is a great way to see Brunel's Bristol. From Queen Square it is a ten-minute walk over the Cliffe Way bridge, past St Mary Redcliffe and along Redcliffe Way across Temple Gate to Temple Meads. There is lots of traffic to negotiate on the way. Alternatively, from Queen Square head towards the bottom of Park Street. Just after the curving Council House building on your left, take the turning down Frog Lane to St George's Road for Brunel House.

Follow round the back of the Council House, over Deanery Road and between the library and cathedral to Anchor Road. Between the modern Explore@Bristol building cross on the Pero's footbridge. A quayside walk takes you round the corner of the Arnolfini Gallery to Prince Street Bridge and the launch site of the *Great Western* on the right immediately after the bridge, next to the M Shed. For the SS *Great Britain*, follow the wharf for another half a mile.

Clifton Suspension Bridge and Docks

See the bridge from two very different perspectives. In Clifton it is best to find a parking space and then walk across it to enjoy the spectacular views. From there the Brunel Lock Entrance is reached by driving down Rownham Hill on the Leigh Woods side of Avon Gorge to join the Brunel Way, leaving it immediately after the river and the big red-brick warehouses. It is usually easy to park nearby.

SOUTH DEVON AND CORNWALL

Saltash is easy to get to and is an ideal central location from which to explore in either direction, for the South Devon Atmospheric Railway or to see some of the Cornish viaducts to the west. (But don't ignore the other viaduct sites that extend across south Cornwall all the way to Penzance and via the branch line down to Falmouth.)

Saltash, Dawlish and Starcross

From Saltash, head up A38 about twenty-two miles to Newton Abbot, which has a nice station dating from the 1920s. The A379 follows the coast through Teignmouth with superb late-Victorian station, and east to Dawlish and Starcross.

Saltash, Liskeard and Viaducts

With so many Cornish viaduct sites to choose from, the A38 westward from Saltash to the Liskeard area covers an interesting selection. Note the Menheniot viaduct north of the road – an example where IKB's stone piers have been capped with brickwork (OS map 201:302608).

Liskeard station (OS map 201: 248637) is interesting with its flying-arch road bridge and fine signal box, and immediately to the south-east is the tall Liskeard viaduct (OS map 201:250635). Continue on the A38 and another mile or so east for the tall Moorswater viaduct (OS map 201:237640). Follow the sign for 'Moorswater Industrial Estate' to get right up to it, although it is difficult to get a bigger view. The original piers stand like a row of broken castle towers beside the newer viaduct.

Telegraph Museum, Porthcurno
This museum near Land's End has an excellent collection of telegraph equipment, including examples of cables and a number of *Great Eastern* items. www.porthcurno.org.uk

OTHER COLLECTIONS AND MUSEUMS

THE SCIENCE MUSEUM

LOCATION: Exhibition Road, South Kensington, London, SW7 2DD.

GETTING THERE: Piccadilly, Circle or District Line to South Kensington.

OPENING TIMES: Daily 10.00–18.00 (closed 24–26 December). Free admission.

You will find models of block-making machinery, the tunnelling shield and all three ships.

INFORMATION: 0870 870 4868 / www.sciencemuseum.org.uk

THE NATIONAL MARITIME MUSEUM

LOCATION: Park Row, Greenwich, London, SE10 9NF.

GETTING THERE: Docklands Light Railway to Cutty Sark station.

OPENING TIMES: Daily 10.00–17.00 (closed 24–26 December). Free admission.

The museum contains a model of SS *Great Britain*, as well as William Parnatt's painting 'Building the Great Eastern'.

INFORMATION: 020 8858 4422 / www.rmg.co.uk

THE NATIONAL PORTRAIT GALLERY

LOCATION: St Martin's Place, off Trafalgar Square, London.

GETTING THERE: Charing Cross or Leicester Square by Tube.
The gallery holds portraits of both Brunels.

OPENING TIMES: Daily 10.00–18.00 (late night Thursdays and Fridays until 21.00). Free admission.

INFORMATION: 020 7312 2463 / www.npg.org.uk

A section of atmospheric railway iron pipe displayed at Swindon Steam Railway Museum.

CHRONOLOGY

1806   9 April: IKB born at Portsea, Portsmouth.

1825   Work starts on the Thames Tunnel.

1827   Thames Tunnel is flooded.

1828   Thames Tunnel abandoned.

1830   IKB wins the Clifton Bridge competition.

       He enrols as special constable during the Bristol Riots.

1833   IKB appointed as engineer for the GWR.

1836   Marries Mary Horsley.

       Foundation stone of Clifton Bridge is laid.

1837   *Great Western* steamship launched in Bristol.

1838   First section of the GWR from Paddington to Maidenhead opens.

1839   Construction of the SS *Great Britain* begins.

1841   The GWR opens from Paddington all the way to Bristol.

1842   Queen Victoria travels on a train for the first time, and she does it on the GWR.

1843   GWR's Swindon locomotive works opens.

       The Thames Tunnel opens.

       SS *Great Britain* launched at Bristol.

1845   Maiden voyage of the SS *Great Britain*.

1846   The SS *Great Britain* runs aground at Dundrum Bay in County Down, Ireland.

1849   The South Devon Railway is completed.

       Windsor branch of the GWR opens.

       Sir Marc Brunel dies.

1854   GWR's Paddington New station opens.

       The Royal Albert Bridge to span the Tamar at Saltash is started.

1855   IKB designs prefab hospitals for the Crimea.

1857   The *Great Western* is broken up.

1858   The *Great Eastern* launched at Millwall.
       15 September: IKB dies.

1864   Clifton Suspension Bridge opened.

1876   SS *Great Britain* is laid up at Birkenhead and offered for sale.

1882   SS *Great Britain* acquired by shipping company for conversion to a sailing ship to
       carry coal from Cardiff to San Francisco.

1886   26 May: The storm-damaged SS *Great Britain* limps into Stanley Harbour on the
       Falkland Islands. Used to store wool.

1892   The end of the broad gauge.

1937   12 April: SS *Great Britain* is towed out of Stanley Harbour and beached at Sparrow
       Cove.

1970   25 March: A salvage party arrives at Sparrow Cove to rescue the SS *Great Britain*.
       19 July: After waiting two weeks for the spring tide, the SS *Great Britain* returns to
       Bristol.

2002   IKB comes second in a BBC poll of Great Britons – only Winston Churchill got
       more votes.

2006   200th anniversary of Brunel's birth.

Detail of a tubular iron girder on the bridge over the South Entrance Lock in Bristol.

BIBLIOGRAPHY

Awdry, Christopher, *Brunel's Broad Gauge Railway* (Oxford Publishing Company, 1992)

Ball, Felicity and Bryan, Tim, *Swindon & the GWR* (Tempus Publishing, 2003)

Barnes, G.W. and Stevens, Thomas, *The History of the Clifton Suspension Bridge* (Clifton Suspension Bridge Trust, 1928)

Beckett, Derrick, *Brunel's Britain* (David & Charles, 1988)

Binding, John, *Brunel's Cornish Viaducts* (Pendragon Books, 1993)

Binding, John, *Brunel's Royal Albert Bridge* (Twelveheads Press, 1977)

Bristol Marketing Board, *Aspects of Railway Architecture* (1985)

Buchanan, Angus, *Brunel – The Life and Times of Isambard Kingdom Brunel* (Hambledon & London, 2001)

Buchanan, R.A. and Williams, M., *Brunel's Bristol* (Redcliffe Press, 1982)

Chapman, W.G. (Ed.), *Track Topics, Great Western Railway* (1935)

Christopher, John, *The Lost Works of Isambard Kingdom Brunel* (Amberley Publishing, 2011)

Christopher, John, *Brunel's Bridges* (Amberley Publishing, 2014)

Christopher, John, *Brunel in Gloucestershire* and *Brunel in Bristol* (Amberley Publishing, 2013)

Christopher, John, *Brunel in London* and *Brunel in Cornwall* (Amberley Publishing, 2014)

Cookson, Gillian, *The Cable – The Wire That Changed The World* (Tempus Publishing, 2003)

Corlett, Ewan, *The Story of Brunel's SS* Great Britain – *The Iron Ship* (Conway Maritime Press, 2002)

Dugan, James, *The Great Iron Ship* (Sutton Publishing, 2003)

Falconer, Jonathan, *What's Left of Brunel* (Dial House, 1995)

Farr, Grahame, *The Steamship* Great Western (Bristol University, 1974)

Fogg, Nicholas, *The Voyages of the* Great Britain – *Life at Sea in the World's First Liner* (Chatham Publishing, 2002)

Fox, Stephen, *The Ocean Railway* (Harper Perennial, 2003)

Gardner, Jack, *Brunel's Didcot* (Runpast Publishing, 1996)

Illustrated London News

Jones, Stephen K., *Brunel in South Wales – Vol. 1, In Trevithick's Tracks* (Tempus Publishing, 2004)

Kentley, Eric, Hudson, Angie and Peto, James (Eds), *Isambard Kingdom Brunel – Recent Works* (Design Museum, 2000)

Leigh, Chris, *GWR Country Stations* (Ian Allan Ltd, 1982)

Lewis, Brian, *Brunel's Timber Bridges and Viaducts* (Ian Allan, 2007)

Lord, John and Southam, Jem, *The Floating Harbour – A Landscape History of Bristol City Docks* (The Redcliffe Press, 1983)

Mackenzie, John M. (Ed.), *The Victorian Vision – Inventing New Britain* (V&A Publications, 2001)

Maggs, Colin G., *The GWR Swindon to Bath Line* (Sutton Publishing, 2003)

Mathewson, Andrew and Laval, Derek, *Brunel's Tunnel … and Where it Led* (Brunel Exhibition at Rotherhithe, 1992)

McIlwain, John, *Clifton Suspension Bridge* (Pitkin Guides, 2000)

Powell, Rob, *Photography and the Making of History – Brunel's Kingdom* (Watershed, 1985)

Rolt, L.T.C., *Isambard Kingdom Brunel* (Longmans Green, 1957)

St John Thomas, David, *The Great Way West* (David & Charles, 1975)

Swindon Chamber for Commerce, *Swindon – Signals from the Past* (1985)

Tames, Richard, *Isambard Kingdom Brunel* (Shire Publications, 2004)

Vaughan, Adrian, *Isambard Kingdom Brunel – Engineering Knight-Errant* (John Murray, 1991)

Opposite: the cutting through Sydney Gardens in Bath.

INDEX